Fundamentals of Temperate Fruits

The Authors

Dr. S. Firoz Hussain working as a faculty in the Division of Fruit Crops, College of Horticulture, Anantharajupeta under Dr YSR Horticultural University. He completed his B.Sc (Hons.) Horticulture, M.Sc (Hort.) Fruit Science from Dr YSR Horticultural University and PhD from ICAR – IIHR, Hesaraghatta, Bengaluru. He qualified ICAR – NET. He has published number of research papers in various national and international journals of repute.

Dr. Matta Lakshminarayana Reddy obtained his B.Sc. (Ag.) from APHU, Hyderabad and M.Sc.(Ag.) major in Horticulture from G.B. Pant University of Agriculture & Technology, Pantnagar. He was awarded Ph.D in Horticulture by UAS, Bangalore. Presently he is working as Dean of Horticulture, Dr. Y.S.R. Horticultural University, Venkataramannagudem. He has published number of research papers in various national and international journals of repute. He also served in teaching as Associate Professor, Professor and Associate Dean (Principal) for more than 10 years and taught UG and PG courses of Horticulture. He is also recipient of Andhra Pradesh State Best Teacher Award 2012 from Government of Andhra Pradesh for his dedicated service in teaching and research and Indira Priyadarshini Award from Health International (NGO) Hyderabad for 2012.

Dr. P. Syam Sundar Reddy is presently working as Associate Professor at College of Horticulture, Dr. Y.S.R. Horticultural University, Anantharajupet, YSR Kadapa Dist, Andhra Pradesh. He has obtained M.Sc., (Agriculture) with specialization in Horticulture and Ph.D (Horticulture) from Acharya N.G. Ranga Agricultural University, Hyderabad. He had five years of experience in Agricultural extension service in Department of Agriculture, Andhra Pradesh and brilliant twelve years of teaching experience in the field of vegetable science. He has the experience of teaching both under graduate, post graduate and Ph.D. students. He has to his credit more than 60 research papers published in various national and international journals.

Fundamentals of Temperate Fruits

S. Firoz Hussain

Matta Lakshminarayana Reddy

P. Syam Sundar Reddy

2020

Daya Publishing House®

A Division of

Astral International Pvt. Ltd.

New Delhi – 110 002

ISBN 9789390371716 (Int. Edition)

Published by : **Daya Publishing House®**
A Division of
Astral International Pvt. Ltd.
– ISO 9001:2015 Certified Company –
4736/23, Ansari Road, Darya Ganj
New Delhi-110 002
Ph. 011-43549197, 23278134
E-mail: info@astralint.com
Website: www.astralint.com

Shri Chiranjiv Choudhary, IFS
Vice Chancellor
Dr Y.S.R. Horticultural University
Andhra Pradesh

FOREWORD

India is bestowed with different types of agro climatic conditions, which includes temperate region ranging from Jammu and Kashmir to Arunachal Pradesh which provides congenial conditions for growing different types of temperate fruit crops. Among the temperate fruits, apple, almond, pear, peach, plum and strawberry are having significant importance. In India, temperate fruit crops are being grown in area of 0.51 million hectares with production of 3.07 million tons. Though the production is high, the productivity has remained low due to problems like alternate bearing, unfruitfulness, fruit drop, replant problem, physiological disorders like bitter pit, sun scald, split pit, Phyllody and albinism and incidence of pests and diseases are the major constraints, confronting the researchers and fruit growers. Thus, there was an ardent need to have a complied literature on this important aspect of fruit production.

The efforts of Dr. Firoz Hussain and Dr. M.L.N. Reddy, in bringing out a comprehensive compilation of scattered information in the form of a valuable book entitled "Temperate Fruit Crops" is praiseworthy. The major feature of this book includes production technologies of different temperate fruits, flowering behaviour, varieties, different species and physiological disorders which have been specifically elaborated and discussed with full justification.

It is hoped that the book would prove to be useful for the students, teachers and researchers and temperate fruit growers of our country and abroad, not only because of its contents, but also because of its simple and understanding language.

CHIRANJIV CHOUDHARY

Preface

In 2018 – 19, India has emerged as a major producer of Horticultural crops (314.87 million tonnes surpassing food grain production 281 million tonnes) and is placed second after China in both fruit and vegetable production. Also horticulture in India with its higher annual growth rate has become a major contributor to growth of Indian Agriculture. It is, therefore, important that horticultural crop production is given more emphasis so that it could sustain the desired growth rate in Agriculture sector across the diverse Agro-climatic conditions.

In India, total fruit production is 92.84 million tonnes, out of which, the production of temperate fruits is only 3.06 million tonnes. The decline in the productivity of temperate fruits is mainly attributed to production problems like alternate bearing tendency, flower induction, fruit drop and incidence of diseases and pests concomitant with physiological disorders in Apple, Pear, Peach, Plum, Almond, Apricot and Walnut etc.

This book covers production technology, rootstocks, training and pruning and physiological disorders of temperate fruit crops. While compiling, I have gone through various research articles, text books and reference books. I acknowledge my humble indebtness to all those authors whose materials are drawn from.

I am very much thankful to my co authors Dr M. Lakshminarayana Reddy (Dean of Horticulture, Dr YSR Horticultural University) and Dr P. Syamsundar Reddy, Associate Professor (College of Horticulture, Anantharajupeta). Apart from them, I am very much thankful and express my sincere gratitude to the editor of this book

Dr. B.N. Srinivasa Murthy, Horticulture Commissioner, DAC and FW, Government of India and Ex Chairman, Coconut Development Board, Kochi, Kerala.

I am very much thankful to Shri Chiranjiv Choudhary, IFS, Vice Chancellor, Dr YSR Horticultural University and Horticulture Commissioner, Government of Andhra Pradesh for forewording this book.

Readers are welcome to point out errors and omissions, if any, and can send their valuable suggestions for improving the quality of book.

S. Firoz Hussain

Matta Lakshminarayana Reddy

P. Syam Sundar Reddy

Contents

Chapter 1

Introduction

The **temperate zones** of the world (latitudes from 35° to the polar circles at about 66.5°, north and south) are where the widest seasonal changes occur, with most climates found in it having balanced influence from both the tropics and the poles. A temperate zone is characterized by freezing temperatures in winter season during which the temperatures may range from 0 to -10°C.

Classification of Temperate Fruits

Fruit Crop	*Botanical Name*	*Chromosome Number*	*Origin*
Apple	*Malus domestica*	2n = 34	Asia Minor (The Caucasus region)
Almond	*Prunus amygdalus*	2n = 16	Central Asian mountain areas
Apricot	*Prunus armeniaca*	2n = 16	Western China
Cherries	a. Sweet cherry: *Prunus avium* b. Sour cherry: *Prunus cerasus*	Sweet cherry: 2n = 16 Sour cherry: 2n = 4 x = 32 (Tetraploid)	South central Europe, Asia Minor
Pear	a. European pear: *Pyrus communis* b. Japanese pear: *Pyrus pyrifolia*	2n = 34	Western China
Peach	*Prunus persica*	2n = 16	China

Fruit Crop	*Botanical Name*	*Chromosome Number*	*Origin*
Plum	a. European plum: *Prunus domestica* b. Japanese plum: *Prunus salicina*	(2n = 6x = 48, Hexaploid)	European plum: Europe Japanese plum: Japan
Walnut	*Juglans regia*	2n = 32	Carpathian mountains in Eastern Europe
Pecan nut	*Carya illinoensis*	2n = 32	North America
Pistachio nut	*Pistacia vera*	2n = 30	Central Asia
Chestnut	*Castanea sativa*	2n = 24	Western Asia

Pome Fruits	*Stone Fruits*	*Tree Nuts*
Apple	Peach	Walnut
Pear	Plum	Pecan nut
Quince	Almond	Pistachio nut
	Apricot	Chestnut
	Cherries	Persimmon

Chapter 2

Apple

Apples are popular because of the many ways that they can be consumed and because of their convenience and durability. Apples may be eaten off the tree or stored for up to a full year. Apples can be processed into sauce, slices, or juice and are favoured for pastries, cakes, tarts, and pies (Downing, 1989). The pulp has been processed into candies (fruit leathers) and used as a source of pectin. The juice can be consumed fresh, either natural or filtered, fermented into alcoholic beverages such as cider or wine, distilled into brandy, or transformed into vinegar. Apples have become the symbol of wholesomeness. "An apple a day keeps the doctor away" is a favourite aphorism and apple pie has become a symbol of goodness along with motherhood. Apple is rich source of carbohydrate and the major acids present in apple are malic and citric acids.

Origin

Apple (*Malus domestica*) is believed to have originated in the Caucasus region near Gilan in Turkestan and domesticated by Greeks and Romans in the Middle East and South–Eastern Europe as a result of their travels and invasions. There are natural forests of apple on the Caucasus Mountains of South–Western Asia and mountains of South Western Asia.

Apple belongs to family **Rosaceae**.

Apple state of India is the **Himachal Pradesh**.

HIMACHAL PRADESH is called **APPLE BOWL of India**.

N=17 is the basic chromosome number. Somatic chromosome number (2n) =34. Some times 51, 68, 85.The majority of cultivated apples are diploids 2n=34, However **triploids 3n=51** and in more recent years **tetraploids (4n=68)** have been observed.

The credit for establishing the apple orchard in Kullu valley of Himachal Pradesh in 1850 goes to **Captain A.A. Lee.**

The cultivated apple has been designated as *Malus domestica*. Some workers prefer to split the species into

Malus domestica - **the cultivated apple**

Malus pumila var paradisiaca - **the paradise apple**

Malus sylvestris – **the wild crab apple**

Malus baccata - **crab apple**

However *M. pumila* is considered to be the **parent of most of our cultivated apples.** The main ancestor of apple is now considered to be *Malus sieversii.*

The first ever exotic introductions were made in 1820 at the Botanical Garden, Ooty and later planted at Kodaikanal. Dried apples are called "**Tsunt Hef**" – an important food item in Kashmir during the winters.

Area: 0.22 million hectares. **Production:** 2.24 million tonnes (NHB Data, 2017-18).

Botany

Deciduous, rarely ever green trees or shrubs, rarely with spiny branches.

1. **LEAVES:** Oval or elliptical to broad ovate, bluntly serrated, becoming glabrous and more or less glossy above and pubescent below.
2. **FLOWERS:** White to pink in cymes, flower of most apple cultivars are epigynous and hermaphrodite, bearing 5sepals, 5 petals, 3whorls of 20 stamens (in central order 10, 5 and 5 stamens) and asyncarpous gynoecium with typical 5 locules.Such a flower develops into a pome. Anthers are yellow, ovary is inferior, and 3-5 celled, style 2-5 connate at base.
3. **FRUIT:** It is a **pome** derived both from the ovary wall and floral tube which is composed of sepals, petal and stamens. This tube is fused with the ovary wall becomes fleshy and ripens with it. The fleshy edible portion is **fleshy thalamus.**

Uses

1. Apple skin contains a unique combination of flavonoids and polyphenols collectively known as phytochemicals.
2. An apple is more efficient than toothbrush for cleaning teeth and gums.
3. Apple can be used in preparation of juice, beverages/**wine like APPLE CIDER,** canned puree, dried apple, canned slices and frozen slices.
4. The pectin in apple is helpful in preventing "**Arteriosclerosis**" a major heart disease in which deposition of cholesterol and hardening of arteries occurs.

5. Apple skin contains five times more vitamin C than flesh.
6. Apple helps the body in absorbing Iron from food.
7. Malic acid in apple dissolves deposits of lime in the body to guard against rheumatism, fibrosis and arthritis.
8. Small apples are richer in vitamin C than larger apples due to higher skin flesh ratio.

Composition of Apple

1. Carbohydrates found in apple consists of sugars (sucrose, glucose, fructose) dextrins, starch, hemicelluloses, cellulose and pectin substances
2. The apple fruits have been found to contain **SORBITAL** (the water soluble alcohol or D-glucose.
3. The acid content, malic and citric acids of fresh fruits helps to develop natural flavours in apple juice.
4. Five quercetin and two phloretin glycosides were isolated from apple skin.

Peak Production Period of Fruit

Himachal Pradesh - July to September.

Jammu and Kashmir - September to October, Uttaranchal - July to mid September.

Diploid Varieties of Apple	*Triploid Varieties of Apple*
Var: Alexander, Golden Delicious, Granny smith, Yellow Newton, Wealthy and Mc-Intosh, Yellow transparent. • Self fruitful • Polinizers are not required • Fruit set is high • 2n=34	Triploids have arisen naturally from fertilization of unreduced gametes. Var: Charlotte, Bisbel Gaint, Stayman Wine Sap, BaldWin, Gravenstein, Rhode Island Greening, Blenheim Orange, Jonagold, Mutsu. • Self un fruitful • Polinizers are required • Fruit set is low • 2n=51

Tetraploid giant sport of apple: "Perrine Transparent".

Climatic Requirements

Temperature

A properly hardened apple tree generally does not suffer any injury even at – 35°C. Most apple varieties have a chilling requirement of **1000 – 1600 hr** at temperatures below 7°C. Apple trees require about 20 – 30 °C temperature during summers. The apple cv. Rome Beauty requires highest chilling requirement **(2700 – 3100 hr)**. Day/Night temperatures maintained at 18/14°C for 4 months, at flower bud initiation produces highest percentage of flower trusses (64.0 per cent) and best quality of flower buds. The newly opened blossoms may be killed at -3.9 to -2.2°C temperature. Temperatures below 4.4°C at blooming not only inhibit the bee activity but hinder the germination of pollen grains. A temperature between **15.5 and 21°C** is ideal for pollen germination. The optimum temperatures for pollination and fertilization are between **20 and 25°C**. The temperatures between 21.1 and 26.7°C are optimum for the germination of pollen grains and growth of pollen tube. A maximum temperature of 22°C and a minimum temperature of 6.2°C results in good fruit set. High temperature **(24.4°C) inhibits Anthocyanin production** while low temperature **(8.1°C during night and 15.8°C during day time) enhances the Anthocyanin production**. Occurrence of low temperature (-5°C) during pre–blooming period resulted in the production of sterile flowers in Golden Delicious Apple.

Soil Requirements

The root growth of apple begins when soil temperature reaches around 7°C. The optimum soil temperature for better vegetative and root growth of apple is around 18°C. Dry matter accumulates in the root at soil temperature between 17 and 22°C. A soil pH of 5.5-6.8 is preferred for apple trees. Liming is essential to correct the pH of acidic soils. Apple trees grow well in wide range of soil types, a deep soil ranging from sandy loam to a sandy clay loam is preferred a minimum depth of 1.6m is desirable the subsoil is probably more important than the upper layer of soil for growth and production of an orchard fruit trees will not tolerate wet soil during growing season.

Different Species

1. ***Malus floribunda* (flowering crab):** It is native of Japan. Flowers are highly ornamental, rose red colour. Fruits are small in size and red in colour. The resistance against apple scab caused by *Venturia inequalis* is contributed by *Vf* genes from *Malus floribunda*.
2. ***M. buccata* (Siberian crab):** It grows wild from Siberia to Manchuria and the Himalayan region. Spreading tree, flowers white and showy. Fruit yellow or red, firm and translucent.
3. ***M. angustifolia* (narrow–leaved crab apple):** It is distributed from Pennsylvania to Tennessee and Florida. Flowers are pink in colour and fruits are small.
4. ***M. coronaria* (American crab apple):** Trees are small, bushy and thorny. Flowers are large with persistent calyx.
5. ***M. ioensis*:** Trees are small, flowers large, fruits are green.
6. ***M. soulardii*:** This is a natural hybrid of *M. sylvestris* and *M. ioensis*.

7. ***M. sylvestris* (Wild crab apple):** Trees are large.
8. ***M. sylvestris* and *M. pumila*:** These are considered as the major ancestral species of modern apple.

Salient Features of *Malus* Species and Bio-types Indigenous to India

Species or Bio-types	*Uses*
Malus baccata var. Himalaica	Used as rootstock in Uttarakhand, as dwarf as M9.
Malus baccata var. Shillong	Easy to propagate clonally, resistance to powdery mildew and as dwarfing as M9
Malus baccata var. Khrot	Easy to propagate clonally, as dwarfing as M9.
Malus baccata var. Giabung	As dwarf as M9.
Malus baccata var. Dhak	Dwarfing
Malus baccata var. Srinagar	Semi dwarf similar to MM 106, resistant to powdery mildew.
Malus baccata var. Rohru	More dwarfing than M9. Resistant to wooly aphids.
Malus sikkimensis var. Schneider	Apomictic seedlings used as rootstock in Sikkim, resistant to powdery mildew and wooly aphids.

Commercial Cultivars of Apple in India

State	*Commercial varieties*
Himachal Pradesh	Starking Delicious, Red Delicious, Rich – a – Red, Tydeman's Early Worcestor, Starkimson, Red Gold, Golden Delicious, Yellow Newton, Red Spur, Top Red, Red Chief, Oregon Spur, Golden Spur, Michael, Schlomit.
Jammu and Kashmir	Red Delicious, Benoni, Cox's Orange Pippin, Ambri, Lal Ambri, Golden Delicious, Jonathan.
Uttarakhand	Early Shanburry, Starking Delicious, Red Delicious, Benoni, Chaubattia Princess, Golden delicious.

Propagation

Apples do not come true to type when propagated by seeds and such the way is vegetative propagation. Vegetative propagation employs grafting or budding on seedlings and clonal rootstocks or raising on their own roots by cutting or layering.

Characteristics of a Desirable Rootstock

1. A good rootstock should make a successful union with the scion cultivar.
2. It should grow relatively fast and yield well.
3. It should provide good anchorage for a tree.
4. It should possess desirable nursery characteristics such as ease of propagation, faster growth and free from excessive branching.

5. It should promote early bearing and heavy production of early scion variety.
6. It should provide some degree of vigour control over the scion variety without adversely affecting productivity or longevity of a tree.
7. It should be resistant to disease such as crown rot and root rot.

Rootstocks

The seeds of crab apple (*Malus buccata*) are used extensively for raising rootstocks.

In Kashmir, wild indigenous crab apple, known as "**Trel**" is used for raising rootstocks.

1. Seedling Rootstocks

1. Seeds of crab apple or any commercial cultivar of apple or used for raising seedling
2. Diploid cultivars like Golden Delicious, Yellow Newton, Wealthy and Mc-Intosh should be preferred as they have better seed content giving better germination
3. Whereas triploid cultivars like Baldwin, Gravenstein, Wine Sap should not be used for this purpose.
4. More than 90 per cent of the apple plants are grafted on seedling rootstocks.
5. In Kashmir wild indigenous crab apple known as **Trel** is used.

2. Clonal Rootstocks

1. Clonal rootstocks of apple are multiplied by stooling as it is the most successful methods
2. The clones are planted at 30 cm distance in rows and 60 cm apart.
3. The mother plant is allowed to grow for one season to get established.
4. The plants are cut back to 13cm from the ground level just before the growth begins.
5. When new growth is about 10cm, the shoots are covered with soil leaving the growing tip, exposed soil is mounded at regular intervals until it is 30-45cm high.
6. At end of the growing season, roots are formed at the base of the covered shoots.
7. Rooted layers are cut off close to the ground level and these should be planted on a deep and fertile soil.
8. The mother stool of 6-10 years old shoot produces 10-12 rooted suckers.

9. Application of 2500ppm IBA at the shoots base after removal of a ring of bark before mounding induced maximum rooting with greater number and length of roots.

Characteristics of Rootstocks

Malling Rootstocks

In 1912, *East Malling Research Station* in England selected and developed a series of 65 rootstocks which were named as Malling stock (M) and were numbered in Arabic numerical. Some of these rootstocks are very dwarfing while others are semi dwarfing and vigorous. All these rootstocks are easily propagated by stooling and trench layering.

A. Dwarfing Rootstocks

M 9

This is the **most widely used rootstock** for apple cultivation.

M 27

It is developed from a cross between **M9and M13** this is the most dwarfing rootstock available in the market. **Ultra dwarfing rootstock for apple**.

M20

1. It originated as rouge in batch of M9 at East Malling.
2. The Cox trees on M20 were found to produce similar tree size as on M27.
3. The disadvantage of M20 rootstock is the emergence of large number of suckers from underground and thus it is unfit for high density orchards.

B. Semi Dwarfing Rootstocks

M2

1. This is the traditional apple rootstock of moderate vigour used in UK and parts Europe.
2. It is susceptible to crown gall and wooly aphid but resistant to crown rot.
3. The leaves are evenly shaped and very close with many lenticels on the stem.
4. Trees on these rootstocks are usually propagated by stooling.
5. Comes to bearing early.

M7

1. These rootstocks produce trees larger than M9.
2. It has a strong root system than M9 and thus needs no staking.

3. Trees on M7 produce fruits early but the rootstock is susceptible to crown gall and crown rot.
4. The shape of the leaves is nearly circular and this can be easily propagated by layering.

M26

1. This is a cross between **M16 and M9** and makes trees smaller than on M7.
2. It is suitable for high density planting.
3. Sometimes it produces even smaller trees than M9.
4. It has good anchorage.

C. Vigorous Rootstocks

M12

1. The trees on this rootstock are very vigorous, requiring wider spacing
2. They come to bearing late.
3. The stocks are, although somewhat resistant to crown gall, sensitive to drought.
4. The root system is well anchored and scion forms a spreading crown.
5. It is difficult to propagate.

M16

1. It is most vigorous rootstocks of Malling series.
2. It is susceptible to collar rot and wooly aphid.
3. Multiplication of these rootstocks is usually done by layering.

M25

1. This rootstock is the result of cross of **Northern spy and M2**.
2. This is very vigorous rootstock makes trees similar to those on M16.
3. It is outstanding in vigour and precocity.

Malling-Merton Rootstocks

The rootstocks in malling series, although had a wide choice for tree vigour have certain disadvantages, particularly their susceptibility to wooly aphids. To overcome this problem the *John Innes Horticulture Institute, Merton* and East Malling Research Station, jointly started breeding programme in 1928 with Malling series and Northern Spy cultivar know to be wooly aphid resistant. The rootstocks which were bred in this series were designated as MM (Malling-Merton) rootstock.

MM104

1. Cross between **Malling 2 and Northern Spy.**
2. The trees on these rootstocks are vigorous and have a well developed root system.
3. The rootstock can be easily propagated by stooling.

MM109

1. Cross between **Malling 2 and Northern Spy.**
2. Trees on MM 109 are semi-dwarf and larger than those on M2.
3. The root system and yield are however, similar to those on M2.
4. It can be propagated easily by stooling.

MM106

1. Cross between **Malling 1 and Northern Spy.**
2. This rootstock produces trees of semi-dwarf nature similar to that of M7 in size and yield.
3. This produces a strong, well anchored tree, has a good number of roots and found to be free from wooly aphid and collar rot.
4. It is well propagated by stooling and is a good rootstock for slight heavy to light soil.

MM111

1. Cross between **Northern Spy and Merton 793.**
2. This rootstock produces semi-dwarfing trees similar to those of M2 but yield about 25 per cent more than M2.
3. It is tolerant to drought.
4. It is easily propagated by layering and hard wood cuttings.

Wooly Apple Aphid Resistant Rootstocks

1. Northern Spy
2. Merton 778
3. Merton 779
4. Merton 789
5. Merton 793
6. *Malus buccata* var. *manchuria*
7. *Malus sikkimensis*
8. *Malus orientale*
9. *Malus purpurea*

Powdery Mildew Resistant Rootstocks

1. *Malus sikkimensis*
2. *Malus prunifolia*
3. *Malus buccata* var. shillong
4. *Malus buccata* var. Srinagar
5. *Malus zumi*

EMLA Series of Rootstocks

EM means East Malling

LA means Long Ashton

EMLA rootstocks are free from virus diseases

EMLA - 9 is much more vigorous than M 9.

Indigenous Species

IARI regional Horticultural Station, Shimla has evaluated rootstocks.

1. *Malus baccata* Rohru – More dwarfing than M9
2. *M. baccata* var. Himalaica – Dwarfing as M9
3. *M. sikkimensis* has a limited use a rootstock due to non availability of virus free bud sticks.

Cultivars Developed or Selected

1. Ambrosia
2. Ariane
3. **Braeburn:** The New Zealand is the world largest producer of this variety.

Sports of Braeburn

Hillwell, Red Braeburn, Lochbuie Braeburn, Joburn Braeburn

1. Cameo
2. Cox's Orange Pippin

Sports of Cox's Orange Pippin

Queen Cox Clone, Cherry Cox, Crimson Cox, Red Cox and Flikweert.

1. Crimson Crisp

Delicious Varieties

1. **Red Delicious:** Widely planted variety of the world.
2. Golden Delicious

3. **Starking Delicious or Royal Delicious**
4. **Skyline Supreme Delicious**
5. **Rich-a-red:** First colour sport developed in 1919.

Colour Strains/Mutants/Sports

Rich-a-red delicious was the first colour sport developed in 1919.

The few others are:

Starking Delicious, Starkimson Delicious, Royal Red, Morgan Spur, Oregon Spur II, Midnight Red, Midnight Red Spur Delicious, Apex, Topred, Imperial Double Red Delicious, Ultrared, Silver Spur.

Description and Identification of Varieties of Apple

Commercial Cultivars

1. **Delicious:** Medium to large in size. Smooth striped skin. Flesh is fine grained greenish white, sweet, juicy with good aroma. Colour is blushed red or golden ripens during September.
2. **Red delicious:** A chance seedling originated in the orchard in USA. A large number of mutants for free growth and skin colour of the fruit have been selected. Cultivar comes to fruiting slightly late. It is ready for picking 140-150 days from full bloom. The fruit is conical with protuberances near the calyx. Skin colour is yellow with sparse red stripes. Flesh is firm sweet and juicy it has moderate alternate bearing and overall productivity is fair. It is self un fruitful maximum storage life is 180 days.
3. **Starking delicious: Bud sport of delicious.** The tree is moderately vigorous with narrow crotch angles and is fairly hardy. It comes to fruiting moderately early; it is ready for picking 130- 140 days from full bloom. Fruits are large conical and skin is yellow with red stripes flesh is firm, juicy and aromatic. It is moderately annual bearer and overall productivity is fair it is self unfruitful maximum storage life is 170 days in cold storage.
4. **Rich-a-red: It is a bud sport of red delicious and has more intense red colour** and is more attractive the yellowish ground colour is overlaid with a bright red wash. There are many tiny dots over the fruit, known as lenticels, the flesh is very juicy, sweet and of excellent flavour it ripens in 130- 140 days from full bloom productivity is fair it is self unfruitful and maximum storage life is 180 days.
5. **Starkrimson delicious: It is a spur type from red delicious**, fruit is conical, and skin has a solid blush, flesh is sweet aromatic and juicy fruits get darker with advance immaturity an increased altitude is a consistent and heavy producer matures in 120-130 days from full bloom.

6. **Tydeman's early Worcester:** It is a cross between Mc Intosh x Worcester pearmine and it was evolved at the East Malling Research station. Tree is vigorous, bearing regularly, fruit is round oblate, skin is solid red and flesh is firm sub acid with pleasant flavour. It has a poor shelf life. It is a good pollinizer for delicious groups. It matures in 120-130 days from full bloom and fetches fairly good prize in market. it suffers from pre harvest fruit drop it is self fruitful.
7. **Red gold:** Thee tree is fairly vigorous. it is used as **pollinizer for delicious group of apples** the tree produces small to medium size apples of dark, dull red colour. The flesh is firm, white, and juicy and sub acid in taste. The heavy bearer but has a tendency towards alternate bearing if allow to over bear. The fruit matures in 140-150 days from full bloom. It has poor storage life.
8. **Ambri: It is the only indigenous variety grown in India,** it had originated in Kashmir. It is an attractive apple with good keeping quality the trees take a long time to come to bearing and crops irregularly. The fruit is irregular to large in size and oblong in shape. It has red streaks over a greenish yellow background; it is a late season apple ripening in a September-October. The pulp is white crisp and sweet. The fruit can last for 4-5 months under ordinary storage in its native place and for 10 months in cold storage. Basically grown in Kashmir it could not be grown successfully in other parts of the country.
9. **Benoni:** The tree is medium grower prolific bearer fruits are medium in size yellow, striped and shaded carmine. Sub acid and ripen in july it is an important early variety of Kashmir.
10. **Irish peach:** The tree is medium, spreading, inclined to droop. Skin is pale yellow, striped carmine with whitish bloom, flesh of the fruit is yellowish, crisp, fine grained, tender, and juicy and sub acid it is an annual bearer and matures in July.
11. **Versified:** This variety known as Maharaja Kashmir. It produces a very vigorous tree which is a prolific bearer. It has a strong biennial bearing habit. Fruit is large sized flat in shape with red stripes, it has white dots in its skin and white dots on its skin and is called white doted red. The fruit is sub acid and slightly sour in taste it ripens late in October.
12. **Cox orange pippin:** It was introduced in India in 1850. It is moderately vigorous grower. Crops well under favourable conditions and good management fruits are medium sized roundish with orange to red flush and bloom. It is susceptible to frost damage powdery mildew and apple canker.
13. **McIntosh:** The tree is vigorous hardy and very productive and matures in August.
14. **Perrine transparent:** Tetraploid

Apomixis

Facultative Apomixis is a characteristic of a number of *Malus* species.

1. *Malus sikkimensis*: Triploid
2. *M. coronaria*
3. *M. hupehensis*
4. *M. torringoides*
5. *M. sargentii*: Tetraploid
6. *M. sieboldii*

Bud Sports Delicious

Spur Type Trees

While locating color sport, growers observed that some of these sports produced a smaller tree bearing numerous fruit spurs. These were mutations of the original red delicious and led to discovery of spur type trees.

1. They have long limbs with numerous fruiting spurs resulting in heavier production.
2. Their potential for higher yield is superior to that of the standard cultivars.
3. These produce the half sized trees (small sized trees) which can be closely planted.
4. Fruits have better colour and are true to type.
5. Internodes and shoots of the spur type are only 80 per cent of standard cultivars.
6. Leaf area and fruit set per 100 blossoming spurs is 20 per cent greater on spur than on standard cultivars.
7. However fruit set by these trees more and requires thinning.
8. They need better soils and require higher doses of fertilizers.
9. They have a great tendency towards alternate bearing.
10. They have upright branches which need to be trained by tying or using tree spreaders.
11. The earliest spur types strain discovered was the Bisbee spur red delicious which is still popular and proven strain.

True Spur Type Varieties

Bisbee (Starkrimson), Red chief, Red spur, Top Red, Ruby red, and Hardee spur, Sturdee spur, Silver spur, Oregon spur and Well spur, Scarlet Spur, Diva, Enterprise, Elstar, *etc.*

Apple Breeding Centers in India

1. Central Institute for Temperate Horticulture (CITH), Srinagar.
2. Sher-e-Kashmir University for Agricultural Sciences and Technology, Srinagar.
3. Dr Y.S. Parmar University of Horticulture and Forestry, Solan.
4. Horticultural Experiment and Training Centre, Ranikhet.

Apple hybrids released from Central Institute for Temperate Horticulture (CITH), Srinagar

1. **Princess Swarnima:** Benoni × Red Delicious
2. **CITH Lodh Apple:** Bud sport of Red Delicious

Apple hybrids released from Sher-e-Kashmir University for Agricultural Sciences and Technology, Srinagar

1. **Lal Ambri:** Red Delicious × Ambri
2. **Sunehari:** Ambri × Golden Delicious
3. **Akbar:** Ambri × Cox's orange pippin
4. **Firdous:** Golden Delicious × Rome Beauty × *Malus floribunda*

Apple hybrids released from Dr Y.S. Parmar University of Horticulture and Forestry, Solan

1. **Ambred:** Red Delicious × Ambri
2. **Ambrich:** Richared × Ambri
3. **Ambstarking:** Starking Delicious × Ambri
4. **Ambroyal:** Starking Delicious × Ambri

Apple hybrids released from Horticultural Experiment and Training Centre, Ranikhet

1. **Chaubattia Anupam:** Early Shanburry × Red delicious
2. **Chaubattia princess:** Early Shanburry × Red delicious

Scab Resistant Varieties

1. Prima: **First scab resistant variety**.
2. Priscilla: **Scab resistant and low chilling variety**
3. Sir prize
4. Macfree
5. Nova easy grow
6. Liberty

7. Coop-12
8. Red free (resistant to apple scab and powdery mildew)
9. Freedom (highly resistant to apple scab, fire blight and powdery mildew)
10. Florina: Sweet variety

Low Chilling Apple Cultivars

1. Tropical beauty
2. Tamar
3. Vered
4. Naomi
5. Maayan
6. Schlomit
7. Anna
8. Michal
9. Winter banana
10. King David
11. Parlin's Beauty
12. Gallia Beauty

Propagation Methods

1. **CHIP BUDDING:** It can be done either in mid-June or during mid-September.
2. **TONGUE GRAFTING:** This is the most successful method of grafting. Tongue grafting is best performed during February-March.
3. **CLEFT GRAFTING: This method of grafting is performed for top working to rejuvenate an apple tree** or to change the variety growing in an orchard for different reasons, mainly when the old variety has lost commercial values or there is a shortage of pollinizer trees.

Alternate Bearing

In apple, some varieties tend to bear heavily in one year, which exhaust the tree and thus it fails to bear a satisfactory crop in the next year. Alternatively, poor fruit set is caused by adverse climatic conditions at the time of flowering lack of pollination and fruiting in the following years and thus the cycle of 'on' and 'off' years are triggered

There exist cultivar differences in apple in respect, to regularity of bearing, which can be categorized in three groups:

Regular Bearing	*Moderate Alternate*	*Strongly Alternate*
Rome Beauty	Delicious	Baldwin
Stayman Winesap	Mc Intosh	Oldenburg
Golden Delicious	Northern Spy	Wealthy
Idared	Rhode Island Greening	York imperial
James grieves	Discovery	Cox's orange
Jonagold	Golden Spur	Pippin
Tydeman's Early Worscestor	Melrose	Laxion's Super
	Starking	Miller's seedling
		Boskoop

It is difficult to outline the precise sequence of distribution in the internal system participating in alteration and its perpetuating mechanism.

1. In the year of a heavy crop (on year), the demand on the carbohydrate supply is such that few flower for the next year. During the following year, when little or no crop is borne, the carbohydrate reserves in the tree build up and an excessive number of fruit buds tend to form.
2. Seeds of growing fruits in apple may also exhibit inhibition effect on flower bud production. Seeds usually increase the number of persisting fruits and reduce self thinning. Both effects are due to enhanced production of growth regulators which induce strong sink activity.
3. There are a large number of reports to suggest that seeds in the growing fruit have inhibitory effect on flower initiation in apple. Gibberellins produced by seeds cause inhibition of flower bud formation upon diffusion resulting in the 'off' year.
4. In an alternate bearing cultivar, diffusible gibberellins have an early peak two weeks after bloom apparently too late to inhibit flower bud formation.

Control of Alternate Bearing

All the control methods are aimed to reduce excess crop in a year or to increase bloom or set in the next year, thus achieving a delicate balance which is essential for preventing alternation.

Methods Suggested to Overcome Alternate Bearing in Apple are

1. Use of regular or fairly regular bearing cultivars.
2. Reduction of flower production in the 'on' year following 'off' year or flower bud development during off year.
3. Thinning of fruit/flower during an early stage in 'on' year.
4. Balancing of reproductive and vegetative growth through pruning and dwarfing rootstocks.

5. The practical way of the control of alternate bearing in apples and other fruits is the use of self compatible cultivar and the use of dwarfing or semi-vigorous rootstocks with response to regularity of bearing.
6. Heavy pruning in winter following 'off' year and less pruning after the 'on' year helps in maintaining the optimal reproductive growth year after year
7. The fruit or flower thinning at early stage during 'on' years and improvement of fruit set with use of plant growth regulators during 'off; year can avert alternate bearing phenomenon
8. Sometimes, the summer pruning is also advantageous in reducing vegetative growth and enhancing flower bud differentiation in 'on' year

Pollination and Fruit Set Problems/Unfruitfulness

Unfruitfulness can be due to lack of balance between growth and fruiting, and lack of flowering and poor fruit set as the result of unfavourable environment

1. It can also be due to heavy cropping leading to exhaustion of fruit bud production and poor crop in the following year.
2. **Self-unfruitful:** Many apple varieties set very little fruit with their own pollen due to self incompatibility (gametophytic) which occur are called self un fruitful.
3. **Self-fruitful:** Some are partially self fruitful, where as others are self fruitful. Even self unfruitful varieties will produce long and regular crops when cross pollinated.
4. **Flower Structure:** In delicious group of cultivar the flowers structure is such that the honey bee can extract the nectar without crawling over the anther and cannot transfer its pollen to stigma, resulting in poor fruit set
5. **Stigmatic Receptivity:** Short duration of stigma-receptivity is also bound to have adverse effect on fruit set.
6. **Climate:** Chilling temperatures, high humidity, heavy frost, snow fall adversely affect the pollination and fertilization of flowers.
7. Knowledge of flowering period and value of variety as cross pollinizer is necessary for fruit growing.
8. Triploid variety such as Baldwin, Stayman Winesap are unsuitable as pollinizer as these produce non viable pollen.

Remedial Measures

1. The varieties Jonathan, Golden delicious, Yellow transparent, Ben Davies, Wagener, Mc-Intosh, Grime golden, Wealthy, York imperial and Rome Beauty have been reported to be very good Polinizers.

2. In Himachal Pradesh Tydeman's Early Worcester, Red Gold and Golden delicious are most desirable Polinizers.
3. Under normal conditions 25 per cent Pollinizers are enough, however in areas where the unfavourable climatic conditions coincide with the time of bloom, at least 33 per cent of Polinizers are desirable.
4. Providing 2 bee hives per acre of orchard and placing of flower bouquets of delicious cultivar trees also facilitate better pollination.

Apple Replant Problem

1. Apple replant problem is a serious and common cause of poor growth and delayed cropping of apple trees planted old orchard sites.
2. It is caused by a complex interaction between biotic (fungi, bacteria, nematodes) and abiotic factors (toxins, nutrition, climatic, soil, water).

Prevention and Control

Prevention of replant problem is much better and more successful than control.

There is very little that can be done to correct replant problems once the trees have been planted. The causes of apple replant problem on different sites are highly variable.

Not all soils respond in the same way to the various pre plant treatments. Thus a treatment that is beneficial in one orchard may have no effect in another.

1. Soil Testing before Planting

Preparation of replanting orchard should start at least one year ahead of planting.

Soil analysis is necessary to determine fertilizer requirements and also use lime to adjust pH prior to planting.

2. Cultural Practices

Special attention to all cultural practices is important to obtain good growth of young trees.

Irrigation and mineral spray requirement are essential.

Good weed control is important to reduce competition for nutrients and water. Trees must also be handled carefully and planted as early as possible.

3. Soil Replacement

Soil replacement with more of new soil, or a well prepared steamed planting soil mixture, can be satisfactory alternatives to treat the planting site.

New top soil should be tested for pH and salinity before use.

It should not come from old apple plantings and should not contain residual herbicides. Soil replacement with a ratio of 1peat to 2 part planting- hole soil can also be a beneficial treatment.

4. Preparation of Site for Fertilizer Treatment

In preparing old orchard sites for replanting, remove as many old roots as possible.

The area to be treated should be cultivated thoroughly before applying the soil chemical or fertilizer treatments.

5. Fertilizers Treatments

Research in the green house using potted apple seedlings has shown that growth can be significantly increased by phosphate fertilizer in 80 per cent of the soils

Ammonium phosphate fertilizer at a rate of 1.0 g/kg of soil should be applied before planting.

Great care must be taken to avoid fertilizer application close to the roots or else burning and death may occur.

Fertigation is an alternative method of applying phosphate to the root zone with less risk of root injury. When using Fertigation lower amount of phosphate can be applied to achieve tree vigour.

6. Soil and Chemical Treatments

Chemical soil fumigation may be used as a preventive treatment for replant disease. Vapam, Busan, Basamid and Telone C - 17 may be used for control of nematodes and soil borne disease in orchard soils prior to planting.

Premature Leaf Fall

As on date, premature leaf fall has become the number one disease problem of apple in H.P. It was just detected in some of the isolated apple orchards in kotkhai fruit belt of H.P state in 1992 and appeared in epidemic form in 1995 in all the major apple growing districts almost at the same time. It causes wide spread defoliation of the apple trees in the midseason itself when the fruits are still on the tree and therefore has been named as pre mature leaf fall. The problem starts in the months of June - July especially in the orchards located at lower elevations and by the August end, only fruits nearing maturity are seen hanging from defoliated branches.

Causal Organism

Detailed studies conducted at scab monitoring and research laboratory, kotkhai in the last decade have led to the conclusion that Marssonine blotch (*Marssonine coronaria*) was the sole cause of the wide spread premature leaf fall problem at apple trees in this state. Since then, its incidence has been reported from neighboring states of Jammu and Kashmir and Uttarakhand.

Symptoms

In H.P symptoms appeared after heavy rains in June to July.

This disease favoured by high rainfall and moderate temperature ranging from 20-22°C during the fruit development stage of apple.

Disease symptoms appear on the leaves and fruits.

Leaf spots usually develop on the upper surface of mature leaves and are 5-10mm in diameter, greyish brown and often tinged purple at the periphery. When lesions are numerous, coalesce and surrounding area turn chlorotic followed by defoliation.

Clear brown circular spots (2-5 mm diameter) also appeared and the surface of the fruits nearing maturity in the defoliated orchard.

Lower portion of the trees was first affected and fruits were commonly seen hanging from the defoliated branches and disease was commonly designated as premature leaf fall.

Initially golden delicious and red gold appeared more susceptible to the disease, however all the commercial cultivars of delicious group were found equally susceptible in due course and slightly and humid locations were found highly favourible.

Control

Repeated applications of broad spectrum fungicide spray provided complete protection of foliage from its attack and leaf shedding.

Protection sprays of Mancozeb (0.3 per cent) COC (0.3 per cent) Zineb (0.3 per cent) provided complete disease control in the field.

Pruning in Apple

Objectives of Pruning

1. To limit tree size and control shape.
2. To thin out competing branches, improve light penetration and spray coverage and facilitate work in the trees.
3. To eliminate weak, unproductive wood.
4. To eliminate or head back excessively vigorous wood.
5. To remove broken or diseased branches.

Initial Pruning

1. Once the apple trees are established, training and pruning should be taken up for future production.
2. Use spur type strains on dwarf rootstock to make training and pruning easier and they produce fruits at an earlier age than the standard types.

3. If one-year-old whip is planted it is usually headed back at the height of 60-90 from the ground.
4. When the new growth is approximately 15cms long, remove all the growth on the bottom up to 45-50cms of the trunk and remove those shoots competing with the central leader.
5. First primary braches are selected with an angle of 45 with the main leader. Three or four shoots that are evenly distributed on trunk should be selected to develop the frame work.

Pruning at Younger Stage

1. If young trees are branched, remove branches those from less than 45 with the main trunk.
2. Eliminate competing leaders by removing less desirable branch.
3. Head back the central leader by 1/3rd in the second year.
4. Make the cut close to a bud that is growing in suitable direction are to a lateral branch
5. Keep pruning to a minimum during the early years to encourage the trees to produce fruiting wood.

Pruning in Bearing Trees

1. Pruning bearing trees is done to maintain a balance between vegetative growth and fruit production.
2. Training is completed in fist 3 years, but in 4th and 5th years trees can be allowed to produce a light crop.
3. Remove week, shaded out wood, diseased or dead wood, water sprouts and root suckers.
4. On the lowest whorl of secondary scaffold tip the terminal shoots rather than cutting back to laterals.

Summer Pruning

1. Summer pruning is advised to remove water sprouts, root suckers and fire blight affected wood
2. It is also done to get specific shape during the initial years
3. Undesired growth should be removed in early summer or after harvest between late August and early September.

Training

Pruning done with a view to giving desired shape to the tree during early growing years is termed as training. Even after proper training the yearly pruning of bearing trees is quite important so that it could support a full crop of large and colored fruits during the subsequent bearing period.

1. Central Leader System

In this system the main stem of the tree is allowed to grow uninterrupted which forms the main central leader with 6-10 strong lateral scaffold branches spread in all directions. This system of training is used to get a large tree and pruning is done during the dormant season. Bearing is confined to top portion of the tree.

2. Open Centre System

In this type of training the central leader is removed about 1m above the ground level. 3-5 well spaced scaffold branches are retained. Secondary scaffold branches are encouraged to develop on the primary scaffold. Both primary and secondary scaffold branches produce fruit bearing laterals. Training by this system keeps the centre open, permitting entry of sunlight throughout the tree. This system facilitates carrying out operations like harvesting and spraying on the trees.

3. Modified Leader System

Standard apple trees are trained to modified leader. In this system the main central leader is allowed to grow for a few years, until 6-8 scaffold branches develop around the central leader. The central leader is cut to form a side lateral, which in due course will grow as a modified leader. In this type of training, the tree develops well spaced limbs with strong crotches. The top being open, allows more sunlight to penetrate deep inside the tree.

4. Cordon

It consists of a single stem with side shoots or fruiting spurs kept short by summer pruning. In this system trees are planted 2-3m apart in rows, 60-90cms apart within the rows. The dwarfing M 9 and semi dwarfing M 26 rootstocks are most suitable for cordons. The trees are usually planted at an angle of about 45 for better exposure to light. Secondary growth arising after summer pruning is cut back to their bases in late September. The trees are supported by 3 rows of wires, stretched tightly between posts. This system offers advantage of early bearing, but is not suitable for vigorous or seedling rootstocks.

5. Dwarf Pyramid

In this system trees are shaped like a Christmas tree, which are planted 1-1.5 m apart in a row and 2-3 m between rows (3-1 m). This system of training is mostly used in England. It is a low headed compact central leader tree with lowest branch at 30-35 cm from the ground. The central leader is not allowed to a height of more than 2m throughout the life of the tree. Lower branches should attain a length of 1m; middle about 60-75cm and, upper branches about 45cm gives the appearance of a pyramid. Rootstocks used are M 26 (dwarf), MM 106 (semi-dwarf).

6. Spindle Bush

This type of training was first developed in Germany. It is a small, conical, central leader tree of only 2m height. Characteristic feature of this method is tying down the lateral shoots in horizontal position with little or no summer pruning. At planting trees are headed back to a height of 75-90cm above the ground. Branches are allowed to grow from central leader at regular intervals choosing wide angled shoots. During growing season select the branches to correct the angle by training them with wires and metal clips. Strong vertical shoots which arise on the main branches are also trained horizontally in the same manner. Higher placed branches are kept shorter for penetrating of light to the lower branches. Rootstock is used as M 9.

7. Espalier

It consists of a central stem from which horizontal branches arise. The distance between rows is 4-6m, Trees are planted 60cm to 1m apart in rows according to the vigor of the rootstock and cultivar and the soil. About 5-6 trees of branches are trained on wire. The unwanted side shoots arising from the horizontal leaders are trimmed and all the laterals from the vertical stem are removed. Annual pruning is done in summer when growth slows down.

High Density Planting in Apple

HDP in Apple were largely established in Europe since 1960's.

Advantages of HDP

1. Early cropping and high yields for a long time.
2. The fruit quality improves with low cost on pruning, spraying and harvesting.
3. HDP induces reduced tree size. As tree size decreases, the shaded unproductive area also decreases.

Characteristics

1. Tress in HDP should have maximum number of fruiting branches and minimum number of structural branches.
2. Prevent upright growth of the trees.
3. Develop horizontal laterals.
4. Space small laterals along the central leader.
5. Develop and maintain fruiting spurs along entire branch as it develops.
6. Develop rigid, strong, self supporting laterals.
7. Maintain fruiting branches in one position.

Methods of Tree Size Control

1. Use of Size Controlling Rootstocks

Most widely used rootstocks are M27, M9, M26, M4, M7, and MM 106. The canopy of smaller apple trees has higher light penetration than those of larger trees. As such, light environment is improved and the orchard efficiency is increased. At Mashobra, 16-year old trees of Vance Delicious on MM 106 at a planting density of 2,222 trees ha^{-1} produced the yield of 75 t ha^{-1}.

Rootstock	*Spacing (m)*	*Trees ha^{-1}*
M9	3 × 2	1666
M26	3 × 3	1111
M7	4 × 3	833
MM 106	5 × 3.5	571
M4, MM111, MM 109	5× 4	500

Recommended Spacing for different Apple Types on M and MM Rootstocks

Type of Variety	*Rootstock*	*Spacing (m)*	*Trees ha^{-1}*
Spur types	MM 111, MM 109	4 × 4	625
Non spur types	MM 106, MM 109	5 × 5	400
Spur types	MM 106, M7	8 × 3	1111
Non spur types	M9	2 × 2	2500

2. Use of Spur Type Varieties

The spur type cultivars are bud sports/mutations of standard commercial varieties mostly of the Delicious group. The trees of spur types are reduced to 50–80 per cent of the standard varieties and are about 20 per cent more spur formation. The introduced spur type varieties which have shown promise are Starkimson, Red spur, Golden spur, Silver spur, Well spur, Oregon spur, Red chief, Hardi spur and Mc Spur.

3. Training and Pruning Methods

In intensive systems of training, pruning is directed to check the vigour of the trees. Widespread use of slender – spindle, HYTEC, North Holland spindle and head and spread system of training for central leader trees has been reported for HDP. Trees trained to spindle bushes had high tree volume, photosynthetic efficiency, fruit set, yield efficiency but yielded smaller, firm fruits with higher TSS and Anthocyanin pigments.

4. Use of Mechanical Devices to Control Free Size

Mechanical devices by laying the use of spreaders and tying down the branches to make them grow from near horizontal to an angle of 45° from main stem to control tree size in intensive system of cultivation.

5. Use of Chemicals

Growth regulators like Daminozide, Alar 85, Ethephon, Cycocel and Paclobutrazol can be used to reduce shoot growth by 1/3 and 1/2. This promotes increased flowering in the subsequent year and may be useful in encouraging earlier commercial fruit production in strongly vegetative, sparingly fruitful young trees. SADH sprayed at a concentration of 3000 ppm during intensive shoot growth in early June at 2000 ppm during petal fall in subsequent years.

Floweing in Apple

Flowers are borne terminally on fruit spurs, laterally on 2-3 year old wood and on tip of the 1-year-old wood. The time between commencement of flower bud differentiation and blossoming may be as long as 8-10 weeks in warmer areas and may be delayed even longer higher elevations.

Induction of Early Flowering

As the young apple tree approaches the stage of fruiting several methods may be employed to induce early development of blossom buds. Measures that will provide vigour control and initiates blossom bud formation are:

1. **Spreading of Branches:** Spreading of apple branches in horizontal position reduces the growth and promotes flower bud formation. The bending of branches of young trees at an angle of 45 from the soil surface reduces the vegetative growth, tree canopy volume, promotes spur formation. Bending restricts the carbohydrate movement and auxin from the upper portion of the limb towards the roots. An accumulation of carbohydrates and slowing down of growth beyond the bend is, therefore, assumed to be favourable to flower bud formation.
2. **Pruning:** Excessive pruning especially dormant tipping of new shoots will induce excessive vegetative growth and inhibits flower bud formation. Hence judicious pruning should be practiced.
3. **Nutrient Control:** Application of N at transitional phase of apple tree is critical. Excessive N application induces more growth and tree remains vegetative. The tree should get sufficient N for optimum growth that would not inhibit flower bud formation. Deficiency of nutrients like Zn, Cu and B is detrimental to flowering in apple.
4. **Rining and Scoring:** Ringing is the removal of thin strip of bark around the trunk or at the base of the main limbs. Whereas scoring consist of making 2 or more parallel cuts to the sap employed on over vigorous trees to check growth and increase flower bud formation.

Disadvantage: It has been observed that scoring can make trees more susceptible to winter injury during the following winter; the scoring and ringing wound on the trunk make the trees more susceptible to severe freeze injury.

5. **Use of Plant Bio-regulators (PBR'S)**: PBR'S are capable of controlling and influencing tree vigour and cropping. The growth regulator Daminozode (Alar) when applied induces flower bud formation. Alar should be applied to those trees that are sufficiently large to produce a crop, but due to excess vigour have failed to develop blossom buds.

NAA is another effective PBR to control growth when used as an additive to paint the pruning cuts.

Paclobutrazol (pp333) is a new chemical which is used to restrict growth and to increase the bloom (Paclobutrazol is a growth retardant and anti-gibberellic in action).

Both Alar and pp333 are used as late July spray and should be preferred over ringing and scoring treatments.

Pollination

Most of the apple cultivars are self-incompatible, but few cultivars like Jonathan, golden delicious and aroma are partially self fertile, pollen germination is good in diploid cultivars than triploid cultivars.

Factors Affecting Fruitfulness in Apple

1. **Defecting Pollen:** Some cultivars produce week sterile or abortive pollen as most of the triploids.
2. **Incompatability:** Pollen of some cultivars is incompatible with certain combinations, *e.g.,* Golden delicious is incompatible with crispin.
3. **Irregular Chromosome Number:** Most apple cultivars are diploids (2n=34). Some important cultivars like Bramleys seedlings, Jupiter are triploids. The triploids are not good pollinizers because of their ploidy levels.
4. **Climatic Conditions**: Low temperatures during flowering (-4 to -2) kills the pistil of the flower. Optimum temperature for pollination and fertilization is 15 - 25°C.
5. **Pollinating Insects:** 2-3 beehives/ha is sufficient for pollination. Bees are effective agents for pollination, but their activity is low below 10°C, during rains and strong winds.

Fruit Drop

The total number of fruits is retained for final harvest decides the total yield/

tree or per unit area. Therefore, it is essential that there must be minimum fruit drop after setting and during the period of fruit development. There are distinct fruit drop in apple

1. **Post Blossom Drop**: This is due to the drop of un pollinated and unfertilized flowers.
2. **June Drop:** This occurs during the summer months and is mainly caused by competition and moisture stress.
3. **Preharvest Fruit Drop:** This is severe in early cultivars such as McIntosh, Tydeman's Early Worcester ranging from 40-60 per cent of crop land. In mid season cultivars like (golden delicious) the range of pre harvest drop varies from 15-20 per cent.
   a. A definite relationship between the auxin content of the seeds and abscission of fruits during various stage of development.
   b. High amount of auxins appear in the seed of 30 days after petal fall, which coincide with the formation of endosperm and cessation of post blossom drop/early drop.
   c. Later a reduction in Hormonal content in seed is observed approximately up to 75 days after petal fall which results in June drop. The hormonal content raises again when the embryo attains the full size.
   d. In the final stage of the fruit growth a rapid decline in seed hormone content is correlated to degeneration of endosperm causing preharvest fruit drop.
   e. These investigations have led to belief that high concentrations of auxins supplied exogenously may inhibit fruit drop.
   f. It has been observed that spraying of NAA (Naphthalene acetic acid) 3 weeks before harvest reduced preharvest fruit drop.
   g. Chemicals such as 2,4,5,-T (2,4,5-Trichloro phenoxy acetic acid) 20 ppm or 2,4,5 TPP (2,4,5 Trichloro phenoxy propionic acid) 15ppm are also quite effective in controlling fruit drop.

Blossom and Fruit Thinning

1. When grown under favourable condition many apple varieties tend to bear heavily than the actual capacity. During an on-year excessive fruiting is an exhaustive process and results in smaller, low quality and unmarketable fruits and limb breakage.
2 During the heavy cropping year, removal of the excess blossom of fruitlets becomes absolutely important to ensure satisfactory development of color, shape, and size of apples remaining on the trees and to encourage flower bud formation for the following year. This may be accomplished by chemicals or hand thinning of blossom or fruits.

a) **Chemical thinning:** The effect of fruit thinning on fruit size is probably is related to leaf/fruit ratio. If this ratio is reduced below 30:1, fruit size is also reduced. Thinning in apple should be done within 30 days after full bloom to improve flower bud production and fruit size this is important because the period of cell division in apple is brief which ends approximately 20 days after full bloom. Thinning in apples during this period stimulates cell division within remaining fruits.

   The most commonly used thinning agents are NAA, NAAm (Naphthalene acetamide) and Carbaryl which improve fruit set effectively in many apple cultivars.

b) **Hand thinning**: During on year, hand thinning of fruits may be employed to improve fruit size and quality of the remaining fruits; however, hand thinning should not be done until the natural fruit drop has been completed. It is a very cumbersome and uneconomical procedure.

Colour Development

Red colour in apple is due to soluble anthocyanin pigments, these arise from a product known as "Chromogen".

Factors which Influence Colour Development

1. **Light:** Ultra violet and blue light affect most red colouration of apple. But these are readily absorbed by dust and moisture in the air. The UV rays are found in higher concentrations at higher elevations and thus at higher elevations, fruits are found to develop better red colour.
2. **Temperature:** High night temperature results in relatively smaller fruits and many cultivars failed to develop characteristic red colour at harvest. Red colour development was found earlier and more intense at cool night temperatures at 11°C. This effect appears due to depletion, of the carbohydrates by respiration, which are required for red pigment synthesis.
3. **Leaf area:** Presence of adequate number of leaves increases sugar content of fruits and help in the development of colour.

   ex- Delicious fruits with 10 leaves (under perfect light exposure) had only 23 per cent of its surface showing solid red colour, where as fruits with 75 leaves showed 58 per cent red colour development and an increase in sugar content by about 10-15 per cent.
4. **Rootstock:** Improvement in fruit colour on a dwarfing rootstock is due to greater exposure of fruits to sun.
5. **Nutrition:** In general level of N which increases fruit yield, usually inhibits proper colour development specially when applied 2-3 months before harvest. High K was found to improve red colour development in red

delicious. The N content of leaf, ranging between 2.2 and 2.4 per cent (on dry weight basis), at midsummer was found optimum for fruit colour in the red delicious and golden delicious varieties in USA.

6. **Position of fruit on the tree:** Fruits with a better appearance often occur on the top and other exposed parts of the tree canopy. Ripened apples in shaded part of the canopy show poor colour but good eating quality. Whereas the fruits on the top of the canopy have a better colour but poor aroma and taste.
7. **Irrigation:** Trees which receive a continuous moderate water supply generally produce good coloured fruits than those grown under stress.
8. **Growth substances:** Different growth substances for improving colour of fruits are 2,4,5-TPP (2,4,5 Trichloro phenoxy propionic acid), 2,4,5-T (2,4,5- Trichloro phenoxy acetic acid), NAA, SADH and ethrel (2-chloro ethyl phosphonic acid).

Maturity Indices for Apple

1. **Seed colour:** Seed of most apple cultivars turn brown at maturity. However seeds may turn brown in certain cultivars several weeks prior to picking maturity and hence this index cannot be relied upon.
2. **Size of the fruits:** Attainment of full size is an indication of fruit maturity. However fruit size depends upon several factors such as crop land, nutrition, availability of moisture so, this is not a completely reliable index.
3. **Skin colour:** Fruit growers generally adopt this index for marketing of their fruits. In red coloured varieties, red colouration is an indication of maturity. Red colouration depends upon several factors such as exposure to light, Nutritional status of tree and Leaf to fruit ratio. Red sports of certain varieties attain red colouration much before ripening, so this is also misleading.
4. **Change in ground colour:** Initially the ground colour of most of the apple varieties is green which changes to light green and later to yellowish green with onset of maturity. When first sign of yellowing appears in the ground colour, the fruit is considered as ready for harvest. In some colour mutants, the entire surface becomes red and no green colour is left for determining maturity. However, this is more reliable maturity index than skin colour.
5. **Flesh firmness:** Optimum range of flesh firmness for different cultivars at maturity is given in the table. Flesh firmness is measured with pressure tester. The firmness is given in kg/pound on the pressure tester.
6. **Starch test:** As the apple fruit approaches maturity, starch content tends to convert into sugar, fruit starts ripening from the core and progresses out ward this is a useful test for apples. Reaction of iodine with starch gives

a dark colour and is used as an indication of starch content or indirectly of maturity.

7. **Soluble solid contents:** A range of total soluble solids and firmness may provide useful information about fruit maturity. However when used alone, soluble solid contents of fruit is an unreliable index of maturity.
8. **Days from full bloom:** Days from full bloom provides fairly reliable information about fruit maturity. Days from full bloom to maturity have been found to remain constant from year to year, regard less of locality and seasonal variations and hence, are a more reliable maturity index than any other index.
9. **Separation from spurs:** As apples approaches maturity, they become more easily separable from the spur. Though ease of separation is a valuable index, yet, it is not fully reliable.

Fruit Maturity

1. Dessert maturity:
   Starch pattern Index of 4.5 – 5.0
2. Shipping maturity:
   Starch pattern Index of 2.0 – 3.0
3. Export maturity:
   Starch pattern Index of 2.0 – 3.5

Starch pattern Index for Delicious apples is 4.0.

Starch pattern Index for Delicious apples during harvesting is 2.8 – 3.5/10.

Grading in Apple

Grade	*Minimum Fruit Diameter (+ 2.5 mm)*
Super Large	85
Extra Large	80
Large	75
Medium	70
Small	65
Extra small	60
Pittoo	55

Harvesting and Postharvest Management

1. Apple is a climacteric fruit, the maturity of the fruit does not coincide with ripening. The fruit usually does not attain fully ripe edible quality on the tree while harvesting. The fruits should be harvested at proper picking maturity to attain proper edible quality at ripening.

2. Picking of immature fruits results in poor quality fruits, lacking flavour and taste which shrivel during storage
3. Over-mature fruits develop soft scald and internal breakdown with poor shelf life.
4. There are several maturity indices which can be adopted for proper fruit harvesting. No single method of maturity index is satisfactory, a combination of maturity indices given below may be better than relying upon any one index alone.

Storage

For most apple cultivars, optimum storage temperature is -1 to 0°C with 90 – 95 per cent relative humidity. The apples held at -1.1°C require 25 per cent longer time to ripen than at 0°C.

Physiological Disorders in Apple

a. Bitter Pit

Symptoms

1. Small brown lesions 2-10 mm diameter (depending on the cultivar) occur as small, dry, brown pockets usually spherical in shape below the skin and also in the flesh of the fruit.
2. The tissue below the skin becomes dark and corky.
3. At harvest or after a period of cold storage the skin develops depressed spots on the surface.
4. These spots generally turn darker and become more sunken the surrounding skin and are fully developed after 1-2 months in storage.
5. Fruit located on vigorous, leafy, upright growing branches have a great potential to develop bitter pit than does fruit that develops from spurs or on horizontal wood near the trees main frame.

Highly susceptible varieties: Granny smith, yellow Newton and Northern Spy.

Less susceptible varieties: Golden Delicious, Mc Intosh.

Causes

1. Unbalanced nitrogen supply or high dose of nitrogen promotes the incidence of bitter pit.
2. A mineral balance in the apple flesh develops with low levels of Calcium and relatively high concentration of potassium and magnesium.
3. Low levels of calcium impair the selective permeability of cell membranes leading to cell injury and necrosis.

Control

Cultural practices that reduce the incidence of bitter pit are:

1. Annual bearing
2. Moderate tree vigour
3. Smaller fruit size
4. Calcium sprays
5. Summer pruning
6. Harvesting mature fruit
7. Early thinning and over thinning can increase bitter pit.
8. Summer sprays of calcium chloride, calcium nitrate or a post harvest dip in calcium solution is recommended.
9. Spray 0.5 per cent calcium chloride (0.5 kg/100 lit of water) starting from first/second week of july at 2 weeks interval for at least 3 times for the control of bitter pit and other calcium related disorders
10. Apple frits can also be enriched in calcium by dipping fruits in 4 per cent calcium chloride (4kg/100 lit of water) solution with 0.03 per cent surfactant for 1- 2 minute, air dried and stored.
11. Boric acid spray 0.1 - 0.2 per cent at 60 and 45 days before harvest increase the calcium status in the apple fruit

b. Cork Spot

Symptoms

1. The symptom is usually the appearance of a small blushed area on the skin of the fruit above the affected brown spot which may be anywhere in the cortex between the skin and core.
2. Cork spot is similar in appearance to bitter pit, but the spots are much deep in the flesh and appear much earlier in the growing season.

Causes

1. Although the exact cause of cork spot is still not fully understood boron and calcium deficiency are occasionally found responsible for development of cork spot

Control

1. Calcium chloride spray was found to reduce the cork spot.
2. Cork spot is more related to the combination of N and Ca rather than Ca alone.

c. Scald

1. Granny smith and Rome beauty, *etc.* are very susceptible.
2. Gala and Fuji are moderately susceptible.

Causes

1. Hot, dry weather before harvest.
2. Harvesting immature fruits.
3. High nitrogen and low calcium concentration in the fruit.
4. Inadequate ventilation in the storage rooms or in packages.

Symptoms

1. Irregular brown patches of dead skin which can become rough when severe, develop within 3-7days upon warming of the fruit following cold storage.
2. The warm temperature do not cause the scald but allow the symptoms to develop from previous injury which occurred during cold storage.
3. Symptoms may be visible in cold storage when injury is severe. In this case, the symptoms intensify upon warming the fruit.
4. Scald is usually not evident until after 3 months of storage.
5. Scald can be more severe on the greener side of the fruit

Physiology

1. The general theory is the alpha-farnesene, a naturally occurring volatile terpene in the apple fruit is oxidized to a variety of products.
2. These oxidation products results in injury to the cell membranes which eventually results in cell death in the outer most cell layers of the fruit
3. Ethylene promotes the formation of alpha-farnesene, and oxygen is required to oxidize alpha-farnesene to conjugated terpenes. Both ethylene scrubbing and low oxygen storage reduces the incidence of storage scald.

Control

1. Scald is controlled, almost everywhere, with anti-oxidant products such as ethoxyquin and diphenyl amine applied after harvest.
2. Both ethylene scrubbers and low oxygen storage reduce the incidence of storage scald.
3. Control of scald in cultivar Granny smith can be achieved by the application of diphenyl amine (2000ppm) and calcium chloride (2-3 per cent) two weeks before harvest.

d. Watercore

Susceptible varieties include Jonathan, Delicious, Granny smith and Fuji.

Symptoms

1. Water core is characterized by water soaked regions in the flesh of the fruit.
2. A preharvest disorder resulting in water soaked in water soaked region in the flesh, hard and glassy in appearance, visible externally when very severe.
3. Water soaked areas are found near the core but may occur in any part of the apple or involve the entire apple.
4. Symptoms often increase rapidly as fruit becomes over mature, but does not increase after harvest.
5. If symptoms are mild to moderate, they may disappear completely in storage.
6. Severely affected fruit may smell and taste fermented
7. The cultivars Delicious, Stayman, Jonathan, Bramleys seedlings, Worcester, Suntan, *etc.* are susceptible to water core.

Causes

1. Large fruit
2. High leaf to fruit ratio
3. High fruit nitrogen and boron
4. Low fruit calcium
5. Excessive thinning
6. High light exposure

Physiology

The water soaked appearance of water core affected fruits results from the accumulation of sorbitol-rich solution in intercellular spaces. Sorbitol is the carbohydrate source translocated into the fruits from the tree. Sorbitol must be converted to fructose by the apple fruit. The reason for accumulation in the intercellular spaces is not known. It is speculated that sorbitol may be translocated to the fruit faster than it can be assimilated perhaps due to an inability of the apple tissue to convert sorbitol to fructose.

Control

1. The most effective way to reduce the incidence of water core is to avoid delayed harvests.
2. As fruit approach harvest maturity, samples of fruit from tree should be cut to look for water care.

3. Lots with moderate to severe water core should not be placed in CA storage but should be marketed quickly.

e. Core Flesh or Brown Core

1. Brown core of apple constitutes a major source of waste in storage with certain cultivars.
2. It is also called core browning or core flush
3. The main characteristic of this disorder is that flush turns brown near the core. Its first appearance is noted as a slight discoloration between the seed cavities
4. This is a low temperature storage problem developing in fruits kept at -1 to 2°C
5. The risk of core flush incidence increases as the level of fruit K increases.

Control

The average incidence of disorder in 1 per cent oxygen was less than in fruit kept in 2 per cent oxygen air.

f. Jonathan Spot

1. Depressed spots of dark brown to black color formed on the skin
2. High level of boron in the fruit and earlier maturation leads to increase in the incidence of Jonathan spot.

Control

Spraying of 0.5- 0.7 per cent of calcium chloride once in few weeks.

Factors Responsible for Low Productivity in Apple

a. Planting of apples mostly belonging to Delicious group:

   Red delicious, Starking delicious, Richared, Starkimson delicious, Red Gold and Golden delicious are being planted.

b. Inadequate pollination.

c. Lack of Suitable rootstocks.

d. Lack of scientific orchard soil management.

e. Lack of proper system of training and pruning.

f. Non-adoption of hi-tech horticultural practices like high density planting, fertigation, biotechnological approaches, *etc.*

g. Lack of irrigation facilities.

h. Old, unproductive and senile orchards.

i. Planting apple trees at unfavourable sites.

j. Production problems like alternate bearing, replant problem, premature leaf fall, incidence of insect – pests (Wooly apple aphid, Sanjose scale, Red mite *etc.*) and diseases like Scab, powdery mildew, root rot, canker *etc.*
k. Climate change
l. Physiological disorders like Bitter pit, Cork spot, Water core and Brown corem, *etc.*

Approaches for Improving the Productivity in Apple

a. Proper Selection of Cultivars

1. Spur types need to be introduced like Red chief, Red spur, Oregon spur and Silver spur, *etc.*
2. Planting of standard colour sports like Top red, Vance delicious, Skyline supreme.
3. Planting of scab resistant cultivars like freedom, liberty, prima, Priscilla, sir prize and Florina, *etc.*

b. Provision of Effective Pollination

1. Red gold, Tydeman's Early Worcester and Golden delicious should be provided in a proportion of 30 per cent.
2. At least 2-3 bee hives per hectare should be kept in the orchard during the flowering season.
3. It would be advantageous to top work 2 – 3 branches on each tree with different pollinizers to ensure cross-pollination and fertilization.

c. Provision of Suitable Rootstocks

1. M 26, M 27, M9, EMLA series, MM 104, MM 106, MM 109 and MM 111.

d. Adoption of Scientific Orchard Soil Management Practices for Conservation of Soil Moisture and Fertility

1. Sod culture
2. Clean cultivation
3. Application of manures and fertilizers on the basis of soil and leaf analysis.

E. Adoption of Suitable Systems of Training and Pruning

F. Adoption of Integrated Nutrient and Disease Management

G. Plantation of Orchards at Suitable Sites.

Chapter 3

Pear

Introduction

Ancient Greek poet HOMER praised pears as one of the **gift of god**. In Asia, China and Japan are the leading pear growing countries. The major improvement occurred in Belgium, France, and England in the 16th, 17th and 18th centuries. In India, the improved pear cultivars were introduced in the later part of the 19th century and the cultivation increased with the success of BARTLETT and GOLA cultivars in hills of Himachal Pradesh and Patharnakh in the sub mountainous region of Punjab. Among the temperate fruits, the pear is next to apple in importance, acreage, introduction and varietal diversity. Because of its tolerance to a wide range of soil and climate conditions, it is grown both in temperate and subtropical conditions. However, the lack of attractive shape and color, and perishable nature of the fruit are serious problems in expanding its cultivation on a very large scale.

Origin and Distribution

The genus *Pyrus* has probably originated in the mountainous regions of Western Asia. VAVILOV (1951) proposed 3 centers of origin (centers of diversity) for the cultivated pears.

1. Chinese Center

It is a primary gene center comprising the regions of North and central china Japan and Korea. The important species which originated in this tract are *P. pyrifolia, P. ussuriensis, P. betulifolia* and *P. calleryana*.

2. Central Asiatic Center

Western Tin Shan, Tajikistan, Uzbekistan, Northwest India and Afghanistan are included in this center. The indigenous species are *P. communis, P. salcifolia, P. salicifolia, P. regeli,* and *P. pashia.*

3. Near Eastern Center

Asia Minor and Caucasus mountains are the tracts. *P. syriaca and P. caucasia* are the other important species found in this center.

Note: (Occidental=Europe and Oriental=Asia)

Commercial oriental cultivars: Asian pears are mainly from *P. pyrifolia* and *P. ussuriensis* and these are all East and North Asian species, *P. pyrifolia* is resistant to blight and drought while *P. ussuriensis* is winter hardy.

Commercial Occidental pears: (*P. communis*) are EUROPEAN and west Asian species *P. communis, P. nivalis, P. salcifolia* and *P. cordata* are endemic to this center Almost all the commercial cultivars of European pear predominate throughout the world except in China and Japan.

Taxonomy

Genus: Pyrus

Family: Rosaceae

The chromosome number X=17, 2n=34, rarely 51 or 68.

Different Species of Pear

P. communis (European pear)

1. The trees are pyramidal in shape, medium tall and spiny when young.
2. The leaves are orbicular-ovate to elliptic, crenate-serrate and glabrous when old.
3. New growth and inflorescence are pubescent.
4. White flowers are borne mostly in corymbs.
5. Fruits vary in shape but mostly pyriform.
6. Calyx persistent.
7. Fleshy pedicels.
8. The pulp is melting and buttery in texture.

P. pyrifolia (Asiatic pear)

1. The tree is tall, vigorous, spreading
2. Leaves are ovate-oblong, dark green, pubescent, elliptical and comparatively large in size.
3. White flowers appear prior to the emergence of leaves.
4. Fruits are mostly round (apple shaped) with a depression at stem end.

5. **Deciduous calyx** very hard and full of grit cells
6. They have crisp texture, a refreshing, sweet taste but lack aroma.

P. pashia

1. It is comparatively medium-sized, open headed, tree locally known as **Mahal or Kainth.**
2. Sometimes stem has stout spines when young.
3. Simple leaves are lanceolate, acuminate, crenate, wooly when young but glabrous and shining above when old.
4. Flowers are white and fragrant.
5. Fruits are globe shaped small dark brown in color, scurfy and covered with raised white dots. They mature during November-December being gritty and unacceptable for table purpose due to **high tannin** content.
6. This species is commonly used as rootstock for commercial cultivars in India.

The cultivars of pear mainly belong to three groups, *viz.* European, Asian and their hybrids:

1. In general the European cultivars are predominant in cultivation as they are of superior fruit quality but susceptible to fire blight.
2. The European cultivars requiring high chilling are Anjou, Bartlett, Bosc, Beurre Hardy, Doyenne du Cornice, Flemish Beauty, Fertility, Starking Delicious, Winter Nelis and Packham's Triumph.
3. In contrast, low chilling cultivars (Asian types) are Keiffer, Le Conte, Gola and Patharnakh.
4. Summer pears are early cultivars with moderate keeping quality while winter pears ripen late and store well for many months.

Composition

1. Water: 83 per cent
2. Carbohydrates:
3. 15 per cent sugars
4. Starch and cellulose are the main carbohydrates found in the pear fruit. In ripe fruits, reducing sugars constitute about 70 per cent of the total carbohydrates present.
5. Fructose (6.5 per cent to 11.2 per cent) Glucose (0.5 per cent to 3.5), Sucrose (0.1 per cent to 2.4 per cent) in the fruit juice.
6. Protein: 0.6 per cent
7. Amino acids: Aspargine: 57 per cent Aspartic acid: 17 per cent Glutamic acid: 8 per cent

8. Organic acids: Malic and Citric acids are most predominant.
9. Vitamins: 60 IU
10. Minerals: Ca-8mg/100gm of fruit pulp, P: 15mg, s: 14mg
11. It is a rich source of pectin and tannins.
12. Characteristic pear aroma is attributed to the presence of **Ethyl and methyl esters of trans-2, cis-4-decadienoic acid in association with n-hexyl acetate**

Uses

1. **Pear** fruits can be used in preparation of several delicious products such as pear jam, pear chutney, pear juice, canned pear and wine making.
2. **Patharnakh** and **keiffer** varieties are considered good for canning
3. Fruits can be consumed by diabetic patients because of low sugar content.
4. It also maintains acid base balance in the body.
5. *Pyrus nivalis* is called as **snow pear** is mainly utilized for making *pear cider* in Europe.

Area and Production

In India, pear is cultivated in an area of 0.04 m.ha with an annual production of 0.312 mt (NHB Data, 2017-18).

Climate and Soil Requirements

1. Pear can be grown in a wide range of climatic conditions, as it can tolerate as low as -26°C temperature when dormant and as high as 45°C during growing period.
2. A large number of pear cultivars require about 1200 hrs below 7°C during winter to complete their chilling requirement to flower and fruit satisfactorily.
3. However, Bartlett needs about 1500hrs. Pear variety Patharnakh needs only 150hr of chilling and can also withstand high temperature and hot winds during summer. The pears having medium chilling requirement are LeConte, Keiffer and pineapple. Hood and Gola perform well in areas experiencing mild winter.
4. Spring frosts are detrimental to pear production and temperature at -3.3°C or below kills the open blossom. Therefore, lowlands should be avoided for its planting. The hail prone areas are also unsuitable, as hail storms affect both plants and fruits.

Soil

Pear grows best in deep, well-drained, fertile, medium textured and relatively more clay soil. **It is more tolerant to wet soils but less tolerant to drought than**

apple. Pears even do well on poorly aerated heavy soil with high water table which is heavy in texture for most of deciduous fruits. A soil depth or about 180cm is ideal for proper root growth and fruit production. Plants growing on deeper soils give about twice the yield than those in shallow soils. A neutral pH range of 6.0-7.5 is desirable because Fe deficiency appears on highly alkaline soil. The highly fertile soils rich in N are not very suitable for pear growing as the incidence of pear psylla and fire blight is more in these soils. An annual rainfall of 100-125cm is considered adequate.

Rootstocks

1. ***P. communis:*** The seedling rootstocks of *P. communis* are compatible with all pear cultivars. They impart vigor to the scion and adapt to a wide range of soil and climate conditions. They are mostly **resistant to oak root fungus** and **pear decline** but are **susceptible to fire blight** and **wooly aphid.**
2. ***P. pyrifolia and P. ussuriensis:*** The susceptibility of these species to pear decline and physiological disorder 'black end' has rendered them unsuitable for using as a rootstock. *Pyrus ussuriensis* is used as a vigorous and winter hardy rootstock in northern Asia and North America.
3. ***P. calleryana:*** This is a blight resistant rootstock, except for its lack of winter hardiness. It produces vigorous trees which are almost free from the problem of 'black end'.
4. ***P. betulifolia:*** This vigorous rootstock is free from wooly aphid, leaf spot and tolerant to alkaline soils and adaptable to wide range of climatic conditions.
5. ***Cydonia oblonga* (Quince)**: Quince is widely accepted **dwarfing rootstock** resistant to wooly aphids and nematodes and susceptible to fire blight and excessive lime and cold.

Quince A is a standard rootstock in Europe and Quince B is semi-vigorous and **Quince C is very dwarfing generally used for high density planting,** early cropping and fruit maturity. Other rootstocks are *Pyrus pashia* (Kainth) *and Pyrus serotin*a.

Many commercial cultivars, *viz.*, Bartlett, Bosc, Eldorado, *etc.*, need an interstock **old home or Hardy** to form a compatible union with quince.

Identification and Description of Varieties of Pear

European Types

Bartlett

1. It is also known as Williams or William's Bartlett or William Bon cherteir.
2. It originated in England.

3. It is the popular commercial cultivar grown all over the world, except in china and Japan.
4. It is a good dessert and canning cultivar.
5. The trees are prolific and regular bearer.
6. The fruit is medium to large, ovate-pyriform, green when picked, turning bright to yellow on ripening.

Max Red Bartlett Pear

1. The cultivar is a bud mutant of Bartlett, and was introduced in 1945.
2. The tree and fruit characters resemble parents; excepting that fruit–skin is dark cranberry red. The tree is tall, spreading and productive and short, and leaves have a red tinge.
3. Fruits mature in about a week later than Bartlett.

Starkrimson Delicious

1. It originated in Michigan, USA, as an all-over red bud sport.
2. It is large in size and one of the most perfectly flavored pears.
3. Its flavor is rich, texture is tender, fine-grained and buttery, with no bitterness or grit cell
4. It is an early cultivar and matures in end of July; 10 days before Bartlett.
5. Fruit size is large and color is rich crimson all over as soon as fruit forms.
6. Color is brilliant red when fully ripe. Flesh is satiny white.

Anjou (Beurre'd Anjou)

1. It is the most important winter pear of French origin, after Bartlett.
2. The tree is vigorous and a regular bearer, but comes into bearing late.
3. It is fairly resistant to cold and fire blight.
4. The fruit is large, bright green, turning greenish to yellow when ripe.
5. The flesh is fine, mildly acidic to sweet with high dessert quality.
6. It has excellent keeping quality and can be stored for about 8 months at -1°C and it is free from storage disorders.

Flemish Beauty

1. It originated as a chance seedling in Belgium.
2. The tree is large, spread in and a high yielder but is susceptible to blight.
3. The fruit is large, obovate, smooth and attractive.
4. The flesh is pure white, very juicy and almost free of grit cells.

5. It has a sweet aromatic flavor and a high keeping quality after ripening.
6. It is self-fertile and is a good pollinizer.

Asian Types

Keiffer

1. It is widely grown pear.
2. The tree is vigorous and productive but self-fruitful.
3. The fruit is medium to large, brownish, gritty, and hard.
4. It is suitable for processing.

LeConte

1. It is a low-chilling cultivar and can be grown in low and mid–hills.
2. The trees of medium growth but bearing are heavy.
3. The fruit is small, round and yellowish green.
4. The flesh is white, juicy and sub-acid.
5. It matures in August.

Gola

1. It is another prolific bearing cultivar.
2. It is grown in low hills.
3. Fruit is roundish but not so large in size.
4. It has an excellent shipping quality.

Patharnakh

1. It is a low-chilling cultivar which is grown in sub-mountainous areas.
2. It can stand high temperatures, hot winds, waterlogging and drought.
3. The tree is vigorous and heavy yielder.
4. Fruits are medium in size with prominent dots.
5. The flesh is juicy, crispy and gritty.
6. It can stand rough handling because of tough skin and a long shelf-life.

Propagation

The pear plants are raised both by budding and grafting on pear suckers and Kainth seedlings. Of the various methods of grafting **tongue and cleft** grafting are the common for pear propagation. When the scion and stock are of equal thickness, tongue grafting can be successfully employed, whereas when the stock is thicker than the scion, cleft grafting can be practiced.

Tongue and cleft grafting are performed during December-January when the plants are still dormant. T-budding is also practiced from May-July. Pears plants can also be propagated through hard-wood cuttings.

Raising of Rootstocks

Raising of Rootstocks from Kainth (*Pyrus pashia*) Seed

Fully mature fruits of Kainth are collected in September. These fruit are dumped for softening and easy extraction of seeds. The seeds need stratification before they are sown in the field or on raised nursery bed. Seeds are sown in sand or sphagnum moss in October-November under the controlled temperature at 4°C to 10°C.

The medium is kept moist and ventilated which facilitates the seeds to germinate within 15-21 days. Under natural conditions germination takes place in about 2 months. These sprouted seeds should be sown in nursery beds at appropriate distance to facilitate propagation work. These may be transplanted in December-January on a second bed at a distance of 20cm x 30cm and a path of 50cm after each two rows is maintained. The seedlings become buddable during April-May or august-September are graftable during December-January.

Raising of Rootstocks from Cuttings (Hardwood Cuttings)

The rootstocks raised from Kainth seed is not uniform in growth and vigor and takes 2 years for the seedlings to become buddable or graftable. Therefore, clonal propagation of Kainth becomes essential to get uniform and true-to-type rootstock. For raising clonal rootstocks, hard wood cuttings have been found to be suitable. Hard wood cuttings of Kainth are prepared from one year old shoots during December-January. These cuttings are treated with **IBA** 100ppm for 24 hr and are placed in moist sand for callusing. The callused cuttings are then planted in nursery rows. These are later budded or grafted. The rootstocks raised from cuttings become graftable, within one year, thus saving on time.

Raising of Rootstocks from Pear Suckers (*Pyrus pyrifolia*)

The root suckers of healthy pear trees are separated during October-December with good root system. These suckers are **tongue grafted** and planted in the nursery at 20cm distance 30cm apart, leaving 50cm space after 2 rows. The root suckers which do not have good root system are planted in nursery rows after heading back to 2-3 buds without grafting. New sprouts come during February-March while becomes buddable during August-September.

Planting Systems

The planting system should provide maximum number of trees per unit area to facilitate the development of each tree and convenience in performing different operations such as manuring, irrigation, training, pruning, spraying, weeding, *etc.* The pear is generally planted by square and quincunx systems.

Square System

In this system plants are planted on each corner of a square. Though the number of plants per unit area is less in this system but it facilitates different operations and winter cropping in the orchard. The average planting distance is 7.5 m from row to row and plant to plant (7.5 m × 7.5 m). In some cases it may be planted ate 5x5 m distance.

Quincunx System

This system is better than the square system and attained by planting an additional tree in the center of each square of permanent trees, plants of peach, plum, guava, rows of pear plant. Care should be taken that the filler trees may not hinder the proper development and training of permanent pear plant. The filler trees should be removed when these start interfering with pear plants.

Planting Distance

Dwarfing rootstock (Quince C) at a distance of 3.5 x 1.0 to 1.25m accommodating 2857 to 2285 plant per hectare.

A pit of 1m x 1m x 1m size is dug at such places and filled with a mixture of soil and well-rotten farmyard manure or compost and 30g Aldrin or BHC dust. Irrigation is given after filling pits to settle down the mixture. At the time of planting, a small hole is dug just big enough to accommodate all the roots. Very long roots can be shortened and plant should remain straight in its position when roots are being covered with soil firmly.

Training and Pruning

Proper training and pruning of pear trees is essential

1. For the development of strong framework
2. To maintain vigour and growth
3. Spread the fruiting area uniformly
4. Secure fruits of good size and quality
5. To encourage regular bearing and
6. To provide convenience of pruning, spraying and harvesting.

Pear trees are usually trained according to **'Modified central leader'** method. In this method 4 or 5 well-spaced limbs are developed during initial 3-4 years and then the leader is headed off. In first year, plants are headed back at 90cm. The lowest branch is allowed to develop at a height of 60cm from the ground level. Four or five primary scaffold branches arising at wider angle, well-spaced, 10-15cm apart and spirally arranged around the tree trunks are selected. Two to three secondary branches are selected on the primary scaffold during second dormant pruning. During subsequent years, training consists of thinning out unwanted branches and cuttings and cutting others to desirable side limbs. The leader should be removed to keep a well-placed, outward growing lateral during last year of training.

Tatura trellis is the latest training system for **high density** planting in pear trees.

In pruning young bearing trees:

1. A certain amount of **thinning out and heading back** of out wards growing laterals are considered adequate.
2. A balance is required to be maintained between fruit production and vegetative growth.
3. Pear bears fruits on spur on **2-year-old wood** and a spur continues to bear for more than 6 years.
4. The limbs with spurs over 6-8 years old needs to be removed in a phased manner.
5. **Vigorous pruning results in fire blight and cork spot.** Therefore, in areas of heavy blight infection pruning should be carried out in a limited manner.
6. Care should also be taken to avoid the injury to the spurs during harvesting fruit.

Pruning of Older Trees

1. The older trees need renewing and invigorating the bearing wood of the trees.
2. This is accomplished by thinning out the un productive secondary scaffold branches entirely.
3. This helps opening up the trees for more light for better color development of fruit. The process of renewal is completed in stages in 3-4 years.
4. The trees again develop new fruiting spurs which bear fruit for another period of 10 years.
5. The pruning of pear should be done in January-early February.

Rejuvenation of Old Pear Orchards

1. Pear trees tend to become less productive at an age of about 20 years. Production on such trees can be restored by rejuvenating them.
2. It can be achieved by heading back the 3-4 main scaffolds to about 15cm during December-January.
3. Rest of the scaffolds should be completely removed. Paint the cut ends with Bordeaux paint.
4. Many sprouts shall emerge on these stubs in March.

5. In May, retain only 1-2 outgoing shoots on each stub with a total of 6-8 shoots per tree.
6. Rejuvenated trees start fruiting in 3rd year and bear commercial crop in the 5th year.

Manures and Fertilization

In hills, the recommended dose for 10 year-old plants is 60-100 kg FYM., 700g N, 350g P_2O_5 and 700g K_2O.

States	*Spacing (m)*	*Age in Year*	*Compost kg/tree*	*(g per tree)*		
				N	*P_2O_5*	*K_2O*
Punjab	7.5 x 7.5	1-3	10-20	50-100	32-96	90-270
		4-6	25-35	200-300	128-192	
		7-9	10-50	350-450	224-228	630-780
		10 and above	50	500	300	900
U.P	5 x 6	1-9	-	25	-	25
		10 and above	-	500	-	325
Tamil Nadu	5 x 5	-	40	600	150	300
Arunachal Pradesh		1-3	20-30	50-150	30-90	30-90
		4-6	35-45	180-300	120-190	120-180
		7 and above	50	350	210	210
H.P	7.5 x 7.5	Full grown trees	100	700	350	700
					(kg/hectare)	
Haryana	7x8	Before bearing	40	12.5	20.0	35.0
		After bearing	40	75.0	60.0	150.0

Irrigation

A major portion of fine roots are distributed in the top 60-90 cm soil depth. A decrease in tree growth is common when the top 90-120 cm soil under the tree is depleted by 50 per cent of available moisture.

1. One irrigation after compost and fertilizers application in last week of January.
2. Another copious irrigation may be given before flowering so that sufficient soil moisture is available throughout flowering.

3. Regular irrigation at 7-10 days interval during April to August as per need.
4. After harvest, the tree may be irrigated at 3-4 weeks interval till they reach the leaf shedding stage.
5. No need of irrigation during December and January.

Inter-cropping

Green gram, Toria and sunflower can be grown in Kharif, while wheat, peas and gram in Rabi season may be intercropped in young orchards. Additional dose of fertilizers should be given to be intercrops. **Peach** can also be planted as fillers in pear plantations.

Weeding

Weeds can be controlled either mechanically by weeding and hoeing or with the use of herbicides. Hexuron 80wp (diuron) @4kg/ha as pre-emergence in the first fortnight of March and Grammaxone 24wsc (paraquat) @ 3litres/ha as post-emergence when leaves are 15-20cm in height is very effective to control wide spectrum weeds.

Flowering and Fruiting

The flowers are epigynous and the ovary is inferior, the type of inflorescence in pear is **corymb.**

The flowering starts in the last week of February and continues upto third week of March. Calyx may be present or rarely deciduous, stamens 15-20. The fruit is borne on spur.

Three types of trees were observed in pears:

1. Self-fruitful
2. Self-unfruitful
3. Partially self-fruitful

Self-fruitful

The pear cultivars grown in the plains of north India are self -fruitful or partially self-fruitful. Patharnakh which occupies about 90 per cent of the area under pear in the Punjab, usually gives commercial crop when planted in solid blocks.

The pear cultivars which are self-fruitful are Beurre hardy and Flemish beauty.

Self-unfruitful

LeConte and **Magness** and are self-unfruitful. In general, planting of **every fourth tree in every fourth row as** pollinizer is adequate in these cultivars.

Partially Self-fruitful

Bartlett and Anjou are virtually self-sterile but they behave has self-fruitful under certain conditions because of production of parthenocarpic fruits.

The extent of self-fruitfulness is determined by the location, season and tree vigor. Even under optimum plant vigor and weather conditions, cross-pollination results in a higher fruit set.

Furthermore, **Parthenocarpic fruits are more prone to preharvest drop than the seeded fruits.**

The cultivars like LeConte, Baggugosha, Smith and Keiffer will give commercial crops, only when at least two of this is planted together. Honey-bees usually prefer flower of other plants to those of pear. This is due to the fact that pear flowers provide a relatively small quantity of nectar which is low in sugar content. In general, planting of every fourth tree in every fourth row as pollinizer is adequate. In addition, 2-3 bee colonies/ha are sufficient for obtaining higher yield.

Fruit – setting in pear start by the end of March. The fruit is born on spurs. The fruit growth in terms of length and diameter follows a pattern of **"sigmoid curve"**

Thinning of Fruits

1. Some pear cultivars, such as Bartlett, hardy, bosc, winter nelis and Anjou **often set very heavy crops requiring thinning** to avoid limb breakage and to improve fruit size and quality thinning is practiced. The degree of thinning, however, depends upon the tree vigor, growing conditions and the fruit set.
2. Thinning is done to obtain the fruits of good size and quality and to stimulate floral bud formation for the next year's crop.
3. 30-40 leaves for fruit in all cultivars should be retained for fruit growth.
4. Therefore, selective removal of smaller and de formed fruits is done regardless of spacing in hand thinning about 3-8 base after petal fall (6 weeks after bloom). Chemical thinning is, however, replacing hand thinning on account of lower cost. **NAA** or **NAAM** at 25 ppm 3-8 days after petal fall has been found very effective for thinning.
5. Recently Ethephon and Paclobutrazol have gained importance in thinning.

Harvesting

1. Pear fruits exhibit a typical climacteric pattern of respiration. Fully matured pear fruits are harvested while still firm and green for canning and market, but for local consumption they are picked at a slightly later stage because fruits hanging on the tree make a considerable gain in size, weight and overall quality.
2. Immature fruits on picking do not attain full size, lacks sweetness and flavor, shrivel quickly in storage and are susceptible to storage scald.

3. In contrast, over mature fruits on picking ripen rather quickly, having unattractive chalky color, gritty in texture, and have a short storage life, and are prone, to core break down, and scald during storage.
4. The dessert quality of Anjou fruit harvested at optimum maturity remains good up to 150 days in storage but deteriorates after 90 days storage in late harvested fruits.
5. Pear fruits are plucked individually by giving a gentle twist rather than direct pull. Two or three pickings at an interval of 3-4 days are better than single picking. In order to avoid bruising at stalk damage and the fruits need careful handling during the harvest.

Maturity Indices

Determination of fruit maturity is very important for successful pear growing which is a definite period between the final stage growth and commencement of ripening. The fruits are picked in this period depending upon their ultimate use. Therefore, a number of maturity indices have been assessed to fix the standards for the optimal maturity.

The factor evaluated determining the maturity of pear fruits are:

Flesh Firmness

Flesh firmness is measured by a **'Pressure tester'**, is the most satisfactory maturity index of pear maturity. A plunger of 0.8cm head is forced into the flesh at 2 or 3 locations on the circumference of a pear fruit recording the reading on scale. The firmness range for Bartlett pear is 10-7.7kg in USA, but 6.7-5.3 kg in Yugoslavia. Hot climate and oriental rootstocks increases the fruit firmness at maturity so its reliability can further be improved if total soluble solids, starch content and heat units are also taken into consideration simultaneously.

Skin Color

Kajiura *et al.* (1975) prepared a fruit ground color chart of Japanese pears for studying the maturity and they have pointed out the calyx end to be the best portion for evaluating the ground color. Lenticels of mature fruits are white which turn brown on the development of cork cells giving an indication that, the fruits if picked, will ripe without shriveling. In case of Bartlett, change in background color from green to whitish color.

Number of Days from Full Bloom

LeConte takes 135-138 days to reach its maturity whereas Patharnakh takes 150 days to reach its maturity (Anjou: 120-150DFFB, Bartlett: 110-133 DFFB, Bosc: 130-145 DFFB).

Other maturity indices are:

Sugar Content

Titrable acidity (malic and citric acid content 0.34-0.45)

Sugar/Acid Ratio

Total soluble solids (11-12.3°Brix)

Starch Test

Corking of the lenticels

Fruit Size

Occurrence of fruit drop

Ease of separation from spur, *etc.*, are reliable maturity indices.

Yield

The yield of fruits varies greatly with the cultivar, age of plant, orchard management and climatic conditions. A well-managed orchard of pear yields **30-35 tones/ha.**

Grading of fruits is very important for better returns. The misshapen, damaged, blemished, scarred fruits should be excluded while grading.

Grades	*Diameter (cm)*
Extra large/extra class	8
Large/Class I	7
Medium/Class II	6.5
Small/Class III	5

The wooden, plastic or cardboard boxes are generally used for packing pears, the fruits should be Packed in layers. The bottom and top of the containers are properly cushioned with newspaper or dry grass for avoiding and bruises to fruits.

The fruits can also be wrapped individually in 10 micron HDPE bags before packing which maintains freshness and improves fruit quality compared with unwrapped fruits.

Postharvest Management

1. Most pear cultivars **ripen** and develop best dessert quality at **15.6°-21°C and 80-85 per cent relative humidity.** High ripening temperatures (above 26°C) may impair flavor and texture of the fruits as they become mealy and fail to ripen.
2. Pears after harvesting are cooled to a core temperature -0.6 to -1.6°C to remove field heat and arrest ripening. Pre-cooling is not necessary if fruits

are to be consumed within a few weeks of harvesting. **Hydro-cooling** reduces the incidence of shriveling and brown core without affecting weight loss or incidence of rot.

Refrigerated Storage

1. It has been observed that the fruits harvested in the last week of July store better. Storage life of pear cultivars ranges from 120 **to 245 days at -1°C at 80 per cent - 90 per cent RH.**
2. At 1-4°C Fruits of early cultivars can be stored for 18-33 days and midseason cultivars stored for 33-70 days and late cultivars can be stored for 52- 158 days.
3. The Patharnakh pear fruits can be successfully stored for 125 days by pre-harvest treatments with calcium nitrate @1.0 and 1.5 per cent one week before harvest.

CA Storage

1. Pear fruit has excellent storage potential in controlled atmospheric storage at -1°C to 0°C

Pear	O_2	CO_2	*Month of Storage*
European pears	1-2.5 per cent	0-0.5 per cent	7-8
Asian pears	2 per cent	1-2 per cent	3-4

The pear fruits particularly those of Patharnakh variety have a long post-harvest life and unique shipping quality and can be transported to distance markets.

Hypobaric Storage/Low Pressure Storage

1. At -1°C-1°C at 60mm Hg pressure fruits can store up to 4-6 months.
2. - 0.6 – 1.6°C arrest ripening.

Physiological Disorders

1. Hard End

Symptoms: When oriental rootstock *pyrus serotina* is used, fruits approaching maturity become hard and black at the blossom end.

Cause: This malady may be attributed unfavorable water relationship between the other plant parts and fruits.

Control: Currently, the use of European pears as rootstock has solved the problem.

2. Pink Calyx (Pink End) Disorder

Symptoms: Pre mature ripening begins with pink coloration near the blossom end. Consequently core break down (brown heart) and softening occur in affected fruits which do not ripen properly.

Cause: The disorder is caused by abnormally cool growing season preceding harvest. Night temperature lower than 7.1°C and day temperature lower than 20°C for few days are sufficient to cause premature ripening.

Control: As soon as the initial symptoms appear, the fruits should be harvested and handled normally.

3. Pear Decline

It is spread by the pest psylla (*Psylla pyricola*).

The three basic syndromes are

1. Quick decline
2. Slow decline
3. Leaf curl

Pear decline disease is caused by Mycoplasma and is transmitted by pear psylla.

Mycoplasma damages the phloem tissues at graft union.

4. Black End

It is due to unfavourable water relationships between the other plant parts and fruits.

Chapter 4

Peach

Introduction

Peach is a temperate fruit and probably the most adapted temperate fruit to warmer climate. Peach cultivation is concentrated between latitudes 30^0 and 45^0 above and below the equator. Most of the peach cultivars require 500 to 1000 or more chilling hours below 7.2^{0}C for normal bloom and fruiting. There are two types of peach cultivars in India (low chilling and high chilling cultivars). The high chilling cultivars are cultivated in the hills of Jammu and Kashmir, Himachal Pradesh and Uttarakhand. Whereas, low chilling varieties are grown in Punjab, Haryana, Parts of Uttar Pradesh and N-E states. The Peach along with its *smooth skin mutant, the nectarine* is a temperature tolerant excellent appearance and quality. It comes in the market early in the season, particularly cultivars grown in warmer regions. Peach can come up well within 2-4 years of planting expanding on a faster pace in many countries and its fruit production is highest among all the fruits.

Origin and Distribution

All evidences proved that the origin of peach is China. Three wild species are commonly found in china are:

Prunus davidiana an ornamental tree growing wild in Northern China and is used as rootstock *P. nira and P. ferangenesis* are indigenous to Tibetan plateau and sinking province, respectively. Besides *Prunus persica*, some other species commonly found in interior Himalayas, are *P. jaquemontii* and *P. undulata* occur commonly. Another species ***Prunus behmi*** is considered as natural hybrid of almond and peach finds its use as rootstock for almond, Peach and Plum.

Peach is now cultivated in most temperate countries particularly in southern Europe. North America, South Africa, Asia and Australia. In India peach cultivation is confined to Nilgiri hills and Uttar Pradesh, Jammu and Kashmir, Himachal Pradesh, Punjab. Low chilling cultivars are grown in submountaneous region of the Himalayas including the north-eastern region of India.

Composition

1. Sugars are found in abundance in the ripe fruits but starch is absent.
2. Sucrose is the major sugar while fructose and glucose are present in equal proportions.
3. The dominant acids are malic, citric and quinic acids.
4. Lactones are responsible for characteristic peach flavour.
5. *Prunasin is the principal glycoside present in the peach and amygdalin is present in the seeds of peach.*
6. Phenolic compounds play a significant role in the quality including colour and flavour development of both fresh and canned peaches. Prominent phenolics found in the mesocarp are chlorogenic acid, epicatechin and cyanidine.
7. The yellow or orange colour of peaches is attributed primarily due to xanthophylls while the red blush skin of some cultivars is due to anthocyanins.

Peach kernel: Fat: 39.9 per cent, Protein: 31.2 per cent, Crude fibre: 14.8 per cent, Sugar: 2.2 per cent and Starch: 3.6 per cent

Differences between Peach and Nectarine

Peach	*Nectarine*
Prunus persica	*Prunus persica var. nucipersica*
Peaches are hairy big sized fruits, possess less aroma and flavour.	The nectarine is a hairless or fuzzless peach, smooth skinned, somewhat like a plum. These are smaller in size, thinner fleshed and may possess stone flavour and aroma.
These have less amounts of organic acids.	These are rich in organic acids.
These can withstand lower temperatures.	These are susceptible to lower temperature
Varieties are Elberta, J.H. Hale, and Babcock.	Varieties are Gold mine, Nectar red, Sun red and Sun rise.

Area and Production

In India, peaches are being cultivated in area of 0.018 m.ha with an annual production of 0.107 mt (NHB Data, 2017-18).

Commercial Cultivars

For table purpose: Cultivars should be yellow fleshed, free stone, regular producer and relatively free from fuzz.

(Elberta, J.H. Hale, Babcock, Cardinal, Dixigem, Redtop, Red Heaven, Candor, Red globe, Baifeng, Shan-e-Punjab, Earligrande, Flordaprice, Parbhat, Partap Para Deluxe).

For dehydration purpose: White-fleshed sweet cultivars having **free stone kernels** are preferred.

For canning purpose: The fruit should have yellow **flesh clingstone,** uniform size and mature evenly and devoid of red colour at pit.

(Golden Queen, Halford, Lovell, Certex, Carolyn Dixon – I, Fortuna, Golden bush, Steuart, Vivian, Veteran, Crawford's Early).

Low chilling cultivars: Shan-e-Punjab, Earligrand, Flordaprince, Parbhat, Partap, Early Amber, Early Grande Flordabelle, Flordagold, Flordaking, Flordasun, Flordaqueen, Rio-Grande, Sharbati and May Gold, Sharbati, Sunred, Sunrise and Babcock.

Nectarine Cultivars

The nectarine cultivar group of peaches has a smooth skin. It is often referred to as a "shaved peach", "fuzzy – less peach" or " shaven peach" due to its lack of fuzz or short hairs. Nectarines have arisen many times from peach trees, often as bud sports. Nectarines can be white or yellow, and clingstone or freestone. The lack of skin fuzz can make nectarine skins appear more reddish.

For example, Punjab Nectarine, May Fire, Silver King, Snow queen, Independence, Annqueen, Charokee, Late Le Grand, Nectared, Sun Grand, Sunlite, Sunred, Sunrise, Sunripe, Red Gold, Summer Bright, Spring Bright and Summer Queen.

Recommended Cultivars of Peach for different States of India

State	*Early Season*	*Mid Season*	*Late Season*
Jammu and Kashmir	Peshwari, Quetta, July Elberta, Saharanpur, Prabhat	J.H. Hale, Alexander, Co - Smith	—
Himachal Pradesh	Alton, World's Earliest, Early White Giant, Red Haven, Stark Red Gold	July Elberta, Kanto – 5, Shimizu Hakuto, Sun Haven	J.H. Hale
Uttarakhand	Flordasun, Early Amber, Shan-e-Punjab, Saharanpur, Prabhat, Florda King	July Elberta, Alexander, Sun Haven, Sharbati.	Para Deluxe, J.H. Hale, Peregrine
Punjab	Partap, Flordasun, Shan-I-Punjab, Florda Red, Sun Red, Early Grande	Sharbati, Khurmani	—

Table Cultivars	*Canning Cultivars*	*Low chilling Cultivars*
1. **J.H. Hale:** It is self unfruitful because of male sterility.	1. Cortex	1. Early Amber
2. Babcock	2. Carolyn	2. Early Grande
3. Cardinal	3. Dixon	3. Flordabella
4. Dixigem	4. Haford	4. Flordagold
5. Red Top	5. Fortuna	5. Flordaking
6. Red heaven	6. Golden bush	6. Flordasun
7. Candor	7. Stuart	7. Flordaqueen
8. Baifeng	8. Vivian	8. Rio Grande
	9. Veteran	9. Sharbati
		10. Ray Gold

Climate

1. In India, peach is mainly grown in mid-hills at a height ranging from 1000-1600 meters above sea level and certain varieties are grown in sub-tropical areas which require only 250-300 hours of chilling during the year for proper flowering and fruiting.
2. Peach trees require chilling period below 7°C for breaking the dormancy and flowering. The chilling requirement varies from 250-850 hours depending upon the variety.
3. The variety like **Redheaven** has a relatively high chilling requirement (800-850 hours) and should not be grown in areas experiencing mild winters.
4. Subtropical climate of Punjab is ideally suited for the cultivation of low chilling peaches. Availability of chilling temperatures upto 250 hours during winter months is enough to break the dormancy of low chilling cultivars (Babcock).
5. Inadequate chilling requirements results in delayed and sporadic foliation deformed and non-viable flower buds and flower bud abscission.
6. A sudden drop from -6.6°C for 2 or 3 days often accompanied by snow following a warm period (15.5-24°C) from November to January can result in severe injury to the trunk crotch and limb of trees, similar injury can also occur in February and March.

The limiting factors in peach cultivation are:

1. Minimum winter temperatures
2. Chilling hours
3. Spring frosts
4. Hail storms

5. High humidity and desiccating winds during summer.
6. However, the most important limiting factor in temperate region is lack of flower bud hardiness either to low dormant temperatures or to frost or freeze conditions during spring nectarines are most susceptible to low temperatures than peaches.
7. The application of potassium, gibberellins between 80-100 ppm in late August and September increase the flower bud hardiness.

Soil

1. Deep sandy loam soil rich in organic matter is best for its successful cultivation. Peaches are **highly susceptible to waterlogging** and prefer perfect drainage.
2. Waterlogging injury is associated with the metabolism of the cyanogenic glycoside, Prunasin, in Prunus spp. In the absence of oxygen, Prunasin get hydrolyzed to hydrogen cyanide, which is autotoxic.
3. Fertile and heavy soils are hazardous as it makes heavy growth and hence results in winter injury.
4. The pH of the soil should be between 5.8 and 6.8.
5. Acidic and saline soils are unfit for peach cultivation.

Propagation

Peaches are commercially propagated by **budding and grafting on seedling peach and apricot rootstocks.**

Seed Propagation of Rootstocks

1. Mature fruits from seedling trees are collected during August-September and the seeds are extracted.
2. Harvesting of fruit at optimum maturity.
3. Removal of fleshy pulp from the pit as soon as possible.
4. Drying of pits slowly.
5. Do not allow the pits to overheat while going through the cleaning and drying processes, the seeds after drying should be stored in dry cool place until sowing time.

Seedling Rootstocks

1. Before sowing, seeds are stratified at 4°-10°C FOR 10-12 weeks in the moist sand. The stratification of peach seeds can also be done under natural conditions.
2. Pre-sowing treatment of seeds with IBA (100 ppm) reduces the stratification period, enhances seed germination and improves the seedling growth.

3. Seeds start germinating in March. The seedlings become buddable in June, if they are grown in fertile soil with full care, they become graftable in the following winter.

Hardwood Cuttings

1. Propagation through cuttings is the easiest and cheapest of all the methods of multiplication.
2. In recent years, it has gained importance in the context of high density planting and clonal propagation of rootstock.
3. Insertion of hard wood cuttings in the field after treating with IBA is the least expensive method of rooting peach cuttings.
4. One-year-old hard wood cuttings are treated with 1000ppm IBA for 10 sec. Then the peach hardwood cuttings kept in rooting medium for 6 weeks provided with a bottom heat of 23°C.

Propagation

1. Peaches are generally propagated by shield or ring budding and occasionally by splice or cleft grafting.
2. In plains **grafting** is performed during **November-January and budding during April-May, June.**
3. In hills, tongue grafting during January-February and T-budding during June and September are performed.

Rootstocks

Seeds for raising rootstocks are obtained from three sources

1. From wild types
2. From commercial cultivars and
3. From special rootstock selections.

Wild Selections

1. *Prunus* spp. such as *P. davidiana.* Rootstocks raised from wild peach seeds lack uniformity and are genetically variable, whereas from the seeds of known cultivars they are more uniform and their performance is predictable.

Commercial Cultivars

P. amygdalus (Almond), *P. Salicinia* (Japanese plums), *P. tomentosa*, *P. besseyi* (sand cherry) are also used as rootstocks for peach in specific situations.

Prunus besseyi and *P. tomentosa* are the dwarfing rootstocks, but these are poorly compatible with peach.

Special Selection

Peach × Almond Hybrids (GF 556, GF 677) (GF=Grande and Ferrade research station, France) are important peach rootstocks in Europe. They are clonally propagated and are useful on alkaline soils due to the resistance to chlorosis.

The rootstock De bale (P. *persica*) **is resistant to drought,** vigorous and adapted to wide range of soils. De bale is resistant to spring, frost, vigorous and self fertile.

Nematodes are most harmful to peach than other pest and diseases. As a result of breeding and selection, **Lovell, Halford, Nemaguard, Nemared, Bailey, Guardian, Shalili, S37, Okinawa and Higma** have been evolved. **Flordaguard** is resistant to root knot nematodes. In cooler regions of India, Wild peach (Kateru), wild apricot (Chulli) and *Prunus mira* (Behmi) are suitable.

Planting

In hilly areas, contour planting system should be adopted if the slope is steep, otherwise terraces should be made.

1. In plains, square system of planting is common. Pits of 1m x 1m x 1m size are dug during September to October with a spacing of 6 x 6m.
2. The pit should be refilled with fertile top soil with 40kg of well rotten FYM.
3. In high density plantation the distance can be reduced to 3mx3m.
4. In Tatura trellis and meadow system, peach is planted at a distance of 5mx1m (200 plants per ha), 2m x 1m (5000 plants per ha) and 1m x 1m (10000 plants per ha) respectively.

Training and Pruning

Among the conventional training systems, *i.e.*, central leader, modified leader and open centre are usually adapted to train peach trees.

In Indian situations, where plenty of sunlight is available, peach trees should be trained in the form of **modified central leader system** with 4-5 scaffold branches.

On the contrary, when sunlight exposure is a limiting factor, vase or open centre system of training is advocated.

Pruning of Young Trees

1. Within 2-3 years of planting in the orchard peach trees from a good frame work consisting of main scaffold and secondary branches.
2. At this stage only light to moderate pruning is required to keep the centre open and the branches properly spaced.
3. Water sprouts and broadly placed branches are also to be restored.

Pruning of Mature Trees

1. Maintaining the balance between vegetative and reproductive wood.
2. Peach bears fruit on one year old wood.
3. For that every year new fruiting branches are to be maintained all around the tree.
4. As per the recommendation pruning should be done in a view that 400 new fruiting branches are to be maintained on the tree assuming a yield of 800 fruits per tree.

Fruit Growth and Development

The development of **peach fruit** follows a typical **double sigmoidal growth curve**. Peach embryos develop vary slowly few weeks after fertilization, whereas pericarp and seed growth vary rapidly, the stone attaining almost full size. The duration of first growth period in cv. Elberta is of about 30-40 days.

1. Second period of retarded growth is due to slow growth of mesocarp (flesh) only but the endocarp and embryo develop rapidly by accumulating dry weight. Lignification of endocarp (pit hardening) is an important feature of this growth period comprising about 4 weeks in Elberta.
2. Third growth period is characterized by increase in size, dry weight and sugar content. Maximum amount of flesh accumulates during this period. This is a 6 week period of Elberta.

Fruit Thinning

1. Thinning is more desirable on mature trees making small annual growth than on young vigorous plants. The heavy bearing cultivars like stark red gold, Pratap, shan-1-punjab, Flordaprince, sharbathi and Early Grande need fruit thinning to minimize the danger of limb breakage.
2. Early cultivar of peach, *viz.*, Pratap and Floridaprince has a much shorter fruit development period. In its case, hand thinning must be completed by the 1st week of April before the fruits show stone hardening.
3. In mid-season cultivars *e.g.*: Shan-e-Punjab, this operation should be completed by the mid-April.
4. In late ripening cultivars like sharbathi should be thinned out by the third and fourth weeks of April, *i.e.*, at the stone hardening stage.
5. In Colins and Red heaven cultivars, appropriate time of **Ethephon application for thinning is 35-45 days after full bloom or when fruits are 10-15mm in diameter. Seed length should be around 12mm.**

Stages of Maturity

Maturation occurs on the plant where as the ripening may takes place before or after harvest. There are five stages of maturity, the fruit passes from one maturity stage to the other in 3-4 days.

1. **Hard**: The fruit is almost of full size, very hard and green. If the fruits are picked at this stage, they will not attain proper eating quality on ripening and astringent in taste.
2. **Firm:** The ground colour begins to change and the fruit yields very slightly to pressure. For a distant market, the fruit can be picked but the fruit quality will still remain inferior on ripening.
3. **Firm-Ripe:** Noticeable change in ground colour occurs and fruit can be pressed in a cupped hand. The fruit is fairly good in taste but has not attained the best eating quality. It can be stored for a long period.
4. **Ripe (Tree Ripe):** This is a fully tree-ripe stage when the fruit yields readily to pressure and is in **prime eating condition of the cultivars.** Fruit can be stored in local market as it can neither be stored nor transported to distant markets.
5. **Soft (soft ripe):** The fruit has lost its firmness and storage life. It will not withstand any transportation.

Manuring and Fertilization

The peach has a relatively high requirement for N and K.

State	*Age of Tree*	*FYM (kg/tree)*	*N*	P_2O_5	K_2O
Himachal Pradesh	6	40	500	250	700
Uttar Pradesh	10	-	300	500	300
Tamil Nadu	-	20	200	1000	1000
Arunachal Pradesh	7	50	350	210	210

Whole quantity of farmyard manure along with P and K is given during December-January. Half of N should be given in spring before flowering and the remaining half a month after flowering later if irrigation facilities are available.

The manures and nitrogenous fertilizers should always be applied by broadcasting evenly in the tree basins which should be sufficiently large and should encompass the entire canopy of the tree. It should be thoroughly mixed in soil by gently raking.

Phosphatic and potassium fertilizers should be applied in trench of 20-25cm width and 10-15m deep made beneath the canopy at a distance of 1-2 m from the main trunk. The trees should be irrigated immediately after the application.

Peach is very susceptible to Fe-deficiency which can be controlled by foliar application of 0.5 per cent -1.0 per cent ferrous sulphate.

Trunk injection of 1 per cent ferrous sulphate or ferric citrate is also beneficial in extreme cases.

Intercropping

Cow pea and French bean can be grown in Kharif, while lentil and peas in Rabi season may be intercropped in young orchards. Weeds can be controlled either mechanically by weeding and hoeing or with the use of herbicides. Simazine @ 3kg/ha as pre-emergence and grammaxone9paraquat)@ 4kg/ha as post emergence when weeds are at 15-20 cm in height.

Flowering

1. Peach is a precocious tree commencing bearing in second or third years after planting during February-March.
2. It bears on the shoots developed in the **one year old wood.**
3. Generally, three buds arise at a node, the side buds being floral and central vegetative.
4. Flowering period in peach is determined by the completion of chilling requirements prevailing temperature during early spring.
5. Low chilling cultivar flordasun blooms in February at Shimla, India. If the temperatures had been abnormally higher, it flowers slightly earlier; it shows that early flowering takes place under high temperatures.

Pollination and Fruit Set

1. Practically, all peach cultivars are self-fertile except J.H. Hale, Elberta Halberta, Chinese cling and Giant.
2. A self-fertile cultivar needs to be planted in double rows alternating with two rows of self-fertile cultivars.
3. Bees in orchards at bloom provide a greater assurance for cross pollination.

Maturity Indices

A large number of maturity indices such as days to maturity, fruit size, firmness, sense of touch, pit discolouration, taste, ground colour, sugar, acidity, starch, sugar to acid ratio, have been assessed by various researches working on different cultivars. Main indices for peach maturity are ground colour change from green to straw colour, flesh firmness and days from full bloom.

Grade	*Fruit Size (mm)*	*No. of Layers*	*No. of Fruits per Layer*
Special	55 and above	3	28-32
Grade I	46-55	4	35-38
Grade II	Below 46	4	38-43

Grade	*Fruit Size (mm)*	*No. of Layers*	*No. of Fruits per Layer*
Extra large	65 and above	28-30	
Grade A	60-65	4	36-40
Grade B	55-60	4	46-50

Storage

1. Peaches have a shorter storage life than majority of other temperate fruits. Cold storage conditions are -0.6-0°C and 90 per cent relative humidity, freestone peaches and nectarines can be kept for two weeks and clingstone for four weeks.
2. In controlled atmosphere storage containing 5 per cent CO_2+1-2 per cent of O_2 at 0°C peaches can be stored upto 42 days.

Physiological Disorders

1. Sunscald

This is the most destructive physiological disorder of peach. It causes severe damage when trunk and tender main scaffold branches are exposed to hot sunlight. Effected portions turn brown in colour and shattering of dead tissue will occur in the areas.

Causes

1. Exposing of branches to hot sunlight
2. Rapid depletion of available soil moisture.

Control

1. Painting of exposed surface with lime paste
2. Shading of branches considerably reduces the incidence, and shading by wrapping straw or hay around the trunk and thicker branches is quite effective in mitigating the problem.

2. Split-Pit and Gumming

In some cultivars in some years, split pit and gumming cause a great loss to the growers because the affected fruits become unmarketable;

1. Splitting occurs at the joint of dorsal and ventral sides of the fruit.
2. Gum exudes from the fruit in large quantity making it unfit for consumption. It may fill the entire pit cavity surrounding the kernel.
3. Seed abortion frequently associated with gumming.

Causes

When heavy rains occur after a long dry period, splitting and gumming have been found accentuated. The exact causes of these problems are yet to be determined.

Chapter 5

Plum

Plum is also known by name 'gage'. The word plum is derived from the word "plume" (old English), which is a modification of Prunum (Latin) or prounmom (Greek). Plum is an important temperate zone fruit crop. It ranks next in importance to the peach so far as the economic importance is concerned. It performs best in low hills and submountaneous regions. Fruits are famous for fresh market and preservation industry.

Origin and Distribution

Watkins (1976) identified 5 centers of origin for plum these include

Europe	European plum	*Prunus domestica* (6X)
Western Asia	Damson plum	*Prunus insititia* (6X)
Western and central Asia	Cherry/Myrobalan plum	*Prunus cerasifera* (2X, 3X, 4X, 6X)
China	Japanese plum	*Prunus salicina* (2X, 4X)
North America	American plum	Prunus americana

Some other American plums are *P. hortulana, P. munsoniana, P. maritiana, P. subcordata, and P. besseyi*

Composition and Uses

1. Plums are rich in sugars and carotenes. Plum pits constitute 2-7 per cent of fruit and the kernels are bitter in taste.
2. A fatty oil can be extracted from the kernels which resembles the almond oil and can be used for lubricating, cooking, illuminating and for hair dressing.

3 The aroma of plum blossoms is due to **benzaldehyde**.
4. Plums are used either as fresh dessert fruits or cooked. They are also canned dried and made into jam, jelly, squash and chutney.
5. Plum fruit is known for its cooling effect and is considered best to overcome the effect of jaundice. Plum is delicious fruit and is generally taken fresh composition of fruit

Plum fruit is also rich in Vit-A and riboflavin. It is a good source of sugars, proteins, carbohydrates and minerals like Ca, P and Fe.

Area and Production

The plum fruits are being grown in area of 0.022 m.ha with production of 0.076 mt (NHB Data, 2017 – 18).

China is the leading country followed by Romania and USA.

In India Himachal Pradesh is the leading producer followed by J&K and UP.

The plum belongs to family Rosaceae and genus *Prunus*. Plum species evolved in the North American center of origin have been placed in the section *Prunocerasus* while plum species evolved in north Asia and Europe have been placed in the section *Euprunus.*

The domestica plums arose by somatic doubling of sterile triploid hybrid of *Prunus cerasifera* (2n = 16) and *Prunus spinosa* (2n = 32). *Prunus domestica* is a hexaploid and has a chromosome number of 2n = 48 whereas, *Prunus salicina* is a diploid (2n = 16).

Prunus spinosa: Black thorn/sloe

Prunus cerasifera: Myrobalan plum

Prunus ussuriensis: Ussurian plum

Prunus nigra: Canada plum

Prunus americana: American plum

Types of Plum

a. European plum
b. Japanese plum
c. Damson plum

a. European Plum/Occidental Group (*Prunus domestica*)

1. It comprises of bulk of the cultivated plums.
2. It is believed to be a hybrid of the. Myrobalan plum and Black thorn/sole.

European plums are categorized into different groups based on shape, size and colour of flesh:

1. Prunes
2. Reine Claude/Green Gage plums
3. Lombard plums
4. Yellow egg

1. Prunes

The distinguishing features of this group of plums are the high sugar content which makes them suitable for drying without removal of pit.

Fruits skin is blue/purple in colour and fruit is oval with bulging ventral side and compressed bilaterally, flesh is firm and thick.

This is commercially the most important group in the eastern world, especially in the North American continent *e.g.*, Italian, Imperial.

2. Reine Claude or Green Gage Plums

These are perhaps hybrids of *Prunus domestica* and *Prunus insititia*

The fruit is greenish yellow, round of good quality, *e.g.* Rien Claude, Yellow Gage, Jefferson, and Washington

3. Lombard Plums

The fruit is purplish red and of good quality, important cultivars of this group are Bradshaw, Pond, and Victoria.

4. Yellow Egg

The fruit is characterized by yellow skin and flesh, *e.g.*, Golden Drop.

b. Japanese/Oriental Plum (*Prunus salicina*)

1. The species is more vigorous, productive, precocious and resistant to diseases than the *Prunus domestica*.
2. It was believed to have originated from china.
3. It was introduced in Japan some 200-400 years ago, from where it got disseminated around the world hence the name Japanese plum.
4. It comes into bloom quite early and thus is a leader for early market. But this feature makes it susceptible to spring frosts.
5. The fruits of Japanese cultivars are usually large and heart shaped often with a pronounced apex which distinguishes them from other plums. A few cultivars are however oblate or round.
6. Fruit set is heavier and consistent one year old wood on spurs.

   e.g. Santa Rosa, Beauty, Maerioposa, Keksey.

c. Damson Plum

1. The plants of these trees are compact and form excellent hedge rows or shelter belts. Their silvery grey blossoms are very ornamental.
2. This is a very hardy species and it has remained largely neglected.
3. This is a small fruited European plum, grows wild through Europe and Western Asia.
4. Plums of these species are commonly divided into Bullaces, Damsons and Mirabells.
   a. **Bullaces:** These are globular green coloured fruits with green pigmentation.
   b. **Damsons:** They are believed o be native of the areas around Damascus. These are oval and purple in colour mostly Used for preservers and jam.
   c. **Mirabells:** The fruits having yellow colour with yellow flesh. They have an excellent flavor. The flavor is retained even after preservation for several years.

Climate and Soil

1. Plums thrive well for sub tropical plains to the temperate high hills.
2. European plums thrive best at 1300-2000m above MSL. They require about 1000-1200hr below 7.2°C during winter to break the rest period.
3. Japanese plum requires 1000-1600m elevation and has a chilling requirement of 700-1000hr.
4. The area with frost free spring and adequate sunshine in summer are most suitable for plum cultivation.

Soil

Plums can grow on a wide range of deep, fertile and well drained loamy soils with pH of 5.5-6 is most suitable the soil should be free from hard pan, water logging and excessive salt. For prunes soil particularly rich in potassium should be preferred as these are heavy potash feeders.

European Plums	*Japanese Plums*
Italian supreme	Beauty
Giant prune	**Santa Rosa:** Leading variety of plum. It is Self-fruitful.
President	Mariposa
Green Gage	
Reine Claude	
Victoria	
Starking Delicious	

Characteristics of Rootstocks

Five different rootstocks are preferred in plums.

a. Myrobalan selections (*Prunus cerasifera*)

1. Most widely used rootstocks.
2. It is vigorous, propagated from seed and grows well in fairly heavy wet soils.
3. It is resistant to drought and crown rot.

   e.g. Myrobalan 29 C, Myrobalan B

b. Marianna: (2n=48)

1. It is a cross between *Prunus cerasifera* and *Prunus munsoniana* wild goose plum.
2. Like the Myrobalan this is a vigorous rootstock, but it is characterized by it resistance to waterlogging and nematodes and is easily propagated through cuttings.

   It originated in Marianna, Texas, USA in 1870's.

   e.g. Marianna 2624

c. *Prunus domestica* Selections

These are not compatible with all plums but used as semi-dwarf rootstocks.

d. *Prunus insititia* Selections/St. Julien Rootstocks

These are dwarfing to semi dwarf rootstocks of plum and these are precocious and are propagated by cuttings, seeds and layering, *e.g.* St. Julien A, St. Julien K

e. Pixy

1. These is one of the recently developed plum rootstocks for which in *vitro* micro propagation techniques have been evolved.
2. The rootstock is **dwarfing**, induced precocity and hastens fruit ripening
3. Pixy has proved compatible with most of the European plum cultivars.

f. American Plum Rootstocks (*Prunus americana*)

1. These rootstocks are useful as moderately dwarfing rootstocks.
2. It is compatible with a wide range of cultivars of European prune, damsons, Japanese plums and native American types.
3. It is propagated from seed and is considered to be useful where winter hardiness is a major consideration, *e.g.* Beach plum (*Prunus maritime*)

   Sand cherries (*Prunus besseyi*)

   Nanking cherry (*Prunus tomentosa*)

g. Romanian Rootstocks

e.g. Rosior Varatic, Corocodus and Otesani 8

h. Other Species and Hybrids

1. Almond and Peach
2. **GF 677:** A selection from *Prunus amygdalus* and *Prunus persica* cross. It is vigorous and resistant to drought but susceptible to waterlogging and root nematode.
3. **GF 557:** A selection from the same cross which produces a vigorous tree resistant to drought, root knot nematode. High pith is however sensitive to waterlogging.
4. Plum is raised on seedling rootstocks of wild apricot called **Zardalu.**
5. In Punjab for heavy wet soils, cuttings of Kabul green gage and for sandy loam soils peach seedlings are recommend as rootstocks.

Seedling Rootstocks

The common rootstocks used for propagation of plum are wild peach wild apricot, *P. behimi* (hybrid of almond and peach). Large propagation of commercial plantations of plum are on peach rootstocks which is suitable for light soil and resistant against layer of moist and sand for a duration of 1-3 months depending on species, at temperature of 3-5°C to break the rest period before they germinate and seedling rootstocks are raised.

Clonal Rootstocks

Plums are propagated on vegetatively perpetuated rootstocks of various vigour groups instead to overcome the unfavourable limitations of soil. Mostly use rootstocks are Myrobalan-B, St. Julien C, Myrobalan 29 can be propagate by stooling and cuttings (750 ppm IBA). The rootstock pixy is increasingly gaining popularity as a dwarf rootstock for use in high density plantation of plum.

Propagation

Plum are propagated vegetatively by budding,and grafting on rootstocks. Grafting is done by tongue method and budding is done either by T- budding or ring budding.

Planting

1. A square planting of 6 m apart in the rows is considered ideal for almost all plum cultivars.
2. Sandy soil- 5.4m x 5.4 m.
3. Deep sandy loam soils and fertile soils - 6.6m x 6.6m or 8.2m x 7.2m.
4. Pits of 1 x 1 x 1m size are dug and filled about a month before planting.

5. Pits should be filled with soil mixed well with fully rotten FYM and filled first 500g SSP, 50g aldrin dust are also added to the pit.
6. Planting is done during December – January.
7. While planting, graft union should be kept at 15-20cm above the ground level to avoid collar rot and scion rotting.

Irrigation

Plum requires adequate amount of water throughout the growing seasons. Generally drought conditions prevail during fruit growth and development therefore irrigation is essential during this period. The peak water requirement in plum is May-June, which coincides with rapid fruit development period.

Manures and Fertilizers

1. Plum requires adequate amount of nutrients for better growth and for quality fruits.
2. The FYM along with full dose of P and K should be applied during December – January.
3. Half dose of N is applied in spring before flowering and remaining half N a month later.
4. Under rain fed conditions, N is applied in single dose, about 15 days before bud break.
5. The N, P and K fertilizers should be applied 30 cms away from the tree trunk and mixed with soil.

Age of the Tree	*FYM (kg)*	*N(g)*	*CAN(g)*	*P_2O_5(g)*	*SSP (g)*	*K_2O (g)*	*MOP (g)*
1	10	70	280	35	220	100	165
2	15	140	560	70	440	200	335
3	20	210	840	105	660	300	500
4	25	280	1120	140	550	400	670
5	30	350	1400	175	1100	500	835
6 and above	35	420	1680	210	1320	600	1000

Training

Training in plum is an important operation in order to

1. Provide strong frame work capable of supporting crop load, and
2. To facilitate easy management practices.
3. Allow proper entry of light and aeration to the plant.

The selection of training system will depend upon the growth habit of plum varieties. Those with spreading growth habit are trained with open center system and for those with upright growth habit modified leader system is considered ideal.

1. In Japanese plums- open center system is more common.
2. Where as in European plums – modified central leader system is the common practice of training.

Open Center System

1. 4-5 main scaffold branches along with few secondaries.
2. The later shoots and water sprouts which are produced, more in Japanese varieties should be removed.

Modified Leader System

1. Equally spaced 3-5 scaffold branches arising with wide angled crotches from the trunk, are selected in the first 2-3 years.
2. The primary scaffold branches are preferably spaced vertically about 15-25 cm apart, spirally around the trunk for balanced distribution load.
3. The growth of central leader is checked or modified at 5m above the ground so that it can no longer grow straight up and growth is diverted outwardly.
4. For convenience of orchard operation the lowest limb is usually kept 60 cm above the ground level.
5. European varieties mostly develop well shaped tree with little pruning is required in Japanese plum to obtain then desired shape.

Pruning

Pruning has to be done based on the bearing behavior of the plum species plum fruits are generally born on one year old wood (Japanese plums) and on spurs of current growth season of the 2-6 years old wood (European plums).

Pruning of Young Trees

1. Maintaining balance between primary and secondary branches and pruning of other branches to get desirable shape in first 2-3 years.
2. Dead and diseased branches and water sprouts are to be removed.

Pruning of Bearing Trees

1. Maintaining balance between primary and secondary scaffolds.
2. The inward growing branches which causes shade to interior branches are to be removed.

In general it is suggested that the bearing plum trees should be pruned in such a way as to make an annual extension of growth about 25 cm is conducive to encourage more spur development with increased productivity regularly.

Pollination and Fruit Set

Fruit set in both European and Japanese plum is a problem.

1. In the plum it is safer to interplant with at least one pollinizer variety while planting with synchronized flowering period for good fruit set.
2. Self-fruitful, self-unfruitful and partially self-fertile varieties were observed in both the plums.
3. Self-fruitful varieties in which 30 per cent of flowers set fruit is considered sufficient for commercial crop.
4. Self-unfruitful varieties set only 1.5 per cent which is in sufficient.
5. Japanese plum varieties are mostly self-unfruitful and require cross. Pollination except var. Santarosa, Beauty which is partially self fruitful self-fruitfulness is not a constantly stable character and may vary from year to year and area to area.

It depends on:

1. Climate
2. Temperature
3. Time of planting

Climate

The cultivar Santarosa, for example considered self-fruitful in India where it gives heavy yield when placed in solid blocks, however, in some northerly located areas of California and Canada, Santarosa requires cross pollination by bees from other cultivars.

Temperature

The germination and development of pollen tube are greatly affected by the prevailing temperature during the bloom period. Pollen tube takes only 4 days to reach the embryo sac at 20-21°C, while at 10-11°C they take 8 days.

Time of Planting

The receptivity of style in plum is maximum upto 6 days, while planting varieties and pollinizers see that the flowering periods should be well synchronized for good fruit set.

To make the pollination of plum varieties more effective, it is

1. Recommended to plant in every third tree in third row with pollinizer.

2. Alternatively two pollinizer rows may be planted after every 6 rows of variety.
3. Provision of 2-3 bee colonies/ha improves pollination and better fruit set.
4. Fruit set in plum could be improved by using plant growth regulators. Single application of GA3 at 50-100ppmwith 2,4,5-TP at 10ppm on the 10th day after full bloom has been reported to increase fruit set.
5. GA_3+ Auxins +N-Diphenyl Urea increases fruit set.
6. Paclobutrazol 500 μg ml^{-1} in autumn followed by 250 μg ml^{-1}spring increases fruit set and increases yield.

The dependable pollinizer varieties are:

1. Vickson
2. Lorade
3. Santa Rosa
4. Red heart
5. Elephant heart

Fruit Thinning

Some plum cultivars tend to over bear, particularly the Japanese plum.

1. Removal of part of the crop enables the remaining crop to receive a greater share of the food manufactured by the leaves.
2. Thinning also results in removing of undesirable fruits and trees, reduces limb breakage.
3. Severe fruit thinning has become a common practice for salicina plums advocated by hand thinning, though the possibility of using mechanical shakers followed by selective, touch up by hand thinning is also recommended.
4. Thinning of plum by use of chemicals is also recommended.

Various Chemicals are Used for Flower and Fruit Thinning in Plum

DNOC: (4, 6 Dinitro Ortho cresol) is used for flower thinning @1000-2500 ppm at full blossoming stage.

CEPA: (2-Chloro Ethyl Phosphonic Acid) or Ethephon can be used for both flower and fruit thinning. For flower thinning it is sprayed 30-40 days after bloom and for fruit thinning applied when seed attains 8-10 mm length.

1. Besides, Carbaryl (sevin) @ 1000 ppm applied at full bloom is also reported to have caused optimum thinning.

Harvesting and Postharvest Management

1. Generally 2 pickings are recommended for plum since all the fruits do not mature evenly on trees at one stage.
2. Fruits are picked without pedicle in such a way that bruising and stem punctures are avoided.
3. Per cooling at 0°C has to be done to arrest the ripening process.

For determining the appropriate time of harvest of fruits, various maturity indices are used:

1. Percent fruit surface colouration
2. Pit discolouration up to 5 per cent
3. Change in the ground colour
4. TSS- total soluble solids 12-16 per cent
5. Firmness 4.1-0.1 per cent kg/cm^2
6. Titrable acidity 0.4-0.8 per cent
7. Days to full bloom-Early varieties -95-105 days

The maturity indices of commercial plum cultivars are given in the following table:

Variety	*Days from Full Bloom*	*Firmness*	*TSS*
Beauty	37-2	5.8-0.45	18-2
Santarosa	94-3	5.9-0.45	16-2
Alubakhara	102-2	5.7-0.45	12-1
Frontier	102-2	6.5-0.45	14-1

Yield

Fully grown plum trees yield 60-70 kg fruit and 10-15 t/ha

Grading

Plums are graded according to their fruit size into 3 grades:

Grade	*Fruit Size (9 mm)*	*No. of Layers*	*No. of Fruits/Layer*	*Box Size (cms)*
Special	42 and above	3	28-32	14.5 x 6.5 x 6.5
Grade1	36 - 42	4	38-43	14.5 x 6.5 x 6.5
Grade 2	Below 36	4	50-56	14.4 x 6.5 x 6.5

They are placed in layers in wooden boxes of 14.5x6.5x6.5 cm size. Firstly boxes are lined with paper and then fruits are packed in layers lined with paper to protect the fruits from bruising during transportation.

Storage

1. Immediately after harvest cooling them to 0°C stops the ripening process in plum for a period of approx 12 days.
2. Dipping plums 4 per cent $CaCl_2$ for 2min increases their storability without any reduction in loss of firmness for 12 days at room temperature
3. Although plums are highly perishable, they can be stored for 2-3 months at 0.2°C

Extensive low temperature breakdown during cold storage of plum these disorders have been observed particularly, when plums are stored at -0.5°C. This low temperature damage comprises of jellying of flesh and is due to damage of mechanism by which insoluble pectin is transformed into soluble pectin.

Low temperature damage can be prevented by:

1. Interrupting cold storage for 2 days holding the fruit ay 18.3°C.
2. Low fruit N content and high Ca and Mg contents at harvest showed a low inci.

Chapter 6

Apricot

Apricots are grown over a wide range of climate from Siberia to North Africa.

In the entire world apricot is cultivated on an area of 4, 85,747 ha with an annual production of 3.83 million tonnes.

Turkey, Iran, Uzbekistan, Italy, France, Morocco, Ukraine, and Japan are top ten apricot producers of the world.

In India apricot is a minor crop cultivated in an area of 4,886 ha producing 16,739 tonnes annually (FAO, 2011). It grows between 900 and 3000 m above mean sea level. Apricot is used as fresh fruit and also dehydrated with or without stone. It is also used for canning and to prepare candy, juice, nectar and puree. Apricot is grown commercially in Himachal Pradesh, Jammu and Kashmir, Uttarakhand and North Eastern hills states.

Apricot is the major fruit crop of cold arid desert region of India, situated between 31° 44′ to 39° 59′ north latitude and 76° 46′ to 78° 41′ east longitude in north-western Himalayas. It includes dry temperature region of India *viz.* Kinnaur and Lahaul Spiti districts of Himachal Pradesh and Leh and Kargil districts of Ladakh region in Jammu and Kashmir state and Kumaon and Garhwal region of Uttarakhand. The climatic conditions of cold arid regions congenial for growing drying type strains of apricot. This fruit crop is grown with minimum cultural care in the tracts, which are otherwise unfit for cultivation of other temperate fruits (Sharma *et al.,* 2005). It is suitable for cultivation in the cold deserts as it can be grown in sandy soils with low moisture and nutrients in soil and extremely low winter temperatures (-35° C). The apricots are highly perishable and around 85 per cent apricots are dried in the dry temperate regions of India for year round availability. The apricot is an attractive delicious and highly nutritious fruit. It is particularly rich source of vitamin-A. It ranks second next to plum. Apricots are grown with minimum cultural care in tracts which are unfit for cultivating other temperate fruit crops. It is a drought resistant, salt tolerant, hardy, plant being rather less susceptible to pests and diseases.

1. Apricot is known as **Chuli** in Ladakh region and the apricots with bitter kernel are called **Khante** meaning bitter, whereas, sweet kernelled ones are called **Nyarmo** meaning sweet.
2. The sweet kernel is eaten without any processing and it is an important source of dietary protein oil and fibre (Targais *et al.,* 2011).
3. Korekar *et al.* (2013) found that in general, the fruit weight of apricot genotypes found in Ladakh regions is low (d" 35.1g) but, they had high tss content (16.1- 20.6).
4. The bitter apricot kernel **"Khante"** is used for body massage and cosmetic purposes which has medicinal properties (Dwivedi and Ram, 2008).
5. Dwivedi and Ram also analysed bitter apricot kernel of Ladakh region and found that the kernel is a rich source of oil (up to 54.21 per cent), protein (17.75-22.56 per cent), carbohydrate (21.16-35.26 per cent), crude fibre (0.84-4.71) and dietary fibre (6.03-22.24 per cent).
6. The lipid profile showed that oleic acid was the primary fatty acid and its content varies from 70.52 to 75.99 per cent in the different samples.
7. In addition, linoleic acid (14.13-22.83 per cent), Arachidic acid (0.08-0.39 per cent) and eicosanoic acid in small quantities have been found.
8. Stearic acid (0.34-1.22 per cent) has been observed as a component saturated fatty acids, but palmitic acid (3.5 -5.04 per cent) and palmitoleic acid (0.56-0.91 per cent) were observed to be present in larger quantities.
9. Low chilling varieties can be grown in hills of Madhya Pradesh and plains of Punjab and Haryana. In Punjab it is grown in Amritsar, Hoshiarpur, Pathankot, Chandigarh and Patiala regions.

Taxonomy

1. Apricot belongs to family Rosaceae, subfamily Prunoideae, tribe pruneae, genus prunus and subgenera prunophora which includes plums, prunes and apricots.
2. Mostly cultivated apricot belongs to species *Prunus armeniaca* L. known as common apricot.
3. However, the generic term apricot include many other apricots *viz.* Ansu apricot, Briancon apricot,Tibetan apricot, Manchurian apricot, Japanese apricot, Siberian apricot and black or purple apricot.
4. Based on random amplified polymorphic DNA (RAPD) apricots are placed within the subgenus *Prunophora* and apart from subgenus Amygdalus.

5. Lingdi and Bartholomew (2003) classified genus Armeniaca into 10 species (***A. vulgaris, A. limeixing, A. sibrica, A. holosericea, A. zhngheensis, A. hypotrichode, A. dasycarpa, A. mandschurica* and *A. mume*).**

Some Important Species of Apricot

1. Alpine apricot (*Prunus brigantiaca* Vill)
2. Ansu apricot (*P. armeniaca var.* Ansu Maxim)
3. Black or purple apricot (*P. dasycarpa*)
4. Common apricot (*P. armeniaca*)
5. Desert apricot (*P. fremontii* S. Wats.)
6. Japanese apricot (*P. mume ISieb and Zucc*)
7. Manchurian apricot (*P. mandshurica*)
8. Siberian apricot (*P. siberica*)
9. Tibetan apricot (*P. holosericea* Batal.)

Apricot Varieties Recommended for Cultivation in India

1. Himachal Pradesh

1. **Mid-hills :** New castle, Early Shipley, Shakarpara
2. **High hills:** 1) Early season: Kaisha and Nugget 2) mid: Royal, Suffaida, Charmagz, Shakarpara and Nari.
3. **Dry temperature zone: -** Charmagz, Suffaida, Shakarpara, Kaisha.

2. Jammu and Kashmir

1. **Ladakh region:** Halman, Rakchey Karpo, Tokpopa, Margulam, Narmu and Khante
2. **Kasmir:** CITH Apricot –I, CITH Apricot –II, CITH Apricot –III Harcot, Turkey, Australian, Charmagz, Rogan and Sharapara.

Promising Apricot Cultivars in India

1. Australian
2. Charmagz: Self-incompatible cultivar which needs pollination with varieties like Turkey.
3. CITH-Apricot- I
4. CITH-Apricot-II
5. CITH-Apricot- III
6. Early Shipley
7. Ema

8. Halman: An important desert cultivar of cold arid zone of India especially, Ladakh region of Jammu and Kashmir.
9. Harcot
10. Kaisha
11. Moorpark
12. Nari
13. New castle
14. Nugget
15. Rakchey Karpo
16. Rogan
17. Royal
18. Shakarpara
19. St. Ambroise
20. Suffaida/Safaida
21. Tilton
22. Tokpopa
23. Wenatchee
24. Wild apricot: It is found in temperate regions of western Himalayas within Himachal Pradesh and Kashmir and locally known as Chulli.

Low Chilling Apricot Cultivars

1. Benazir
2. Chaubattia Alankar: Hybrid between **Kaisha x Charmagz**
3. Chaubattia Madhu: Hybrid between **Turkey x Charmagz**
4. Chaubattia Kesari: Hybrid between **St. Ambroise x Charmagz**

Origin and Distribution

To be very precise, its centre of origin in northeast China between the town Kan-Techno and Russian border. The species name armeniaca, indicates its introduction to Italy and Greece by Armenian traders.

Composition and Uses

Moisture	85.3 per cent
Protein	1.0 per cent
Fat	0.3 per cent
Fibre	1.1 per cent
Carbohydrates	11.6 per cent

Mineral matter	0.7 per cent
Calcium	20 mg
Phosphorus	25 mg
Iron	2.2mg
Vitamin-A	3600LU
Thiamine (Vit- B_1)	0.04mg
Riboflavin (Vit -B_2)	0.13mg
Nicotinic acid	0.6 mg
Ascorbic acid	6mg

Reducing sugars of the fruit increase steadily with its development, while sucrose appears at the fully ripe stage.

The pulp contains about 1 per cent pectic substances as calcium pectate.

The apricot fruit has got a distinct pleasant aroma imparted to it by many volatile constituents. It is because of benzaldehyde, linalool, 4-terpinol, alpha-terpinol, and 2-phenylethanol.

Uses

1. A tree ripen apricot is an excellent dessert fruit. Because of its perishable nature, the fruit is canned, candied, frozen and dried.
2. Apricot is the rich source of Vitamin A.
3. The fruit is also processed into a number of products, such as jam, nectar papad and apricot leather.
4. The fruit pulp is thinly spread, dehydrated and pressed to make papad.
5. **Apricot leather (kamaradin)** is similarly prepared by drying the pulp after treating it with sulphur dioxide as preservative and mixing the same with sucrose and glucose syrup.
6. Another popular beverage, apricot nectar is made by converting the ripe fruits into puree and mixing it with sugar syrup and citric acid.
7. Apricot is also consumed as dry fruit.
8. The apricot seed represents about 15 per cent of the fruit and kernel represents about 34 per cent of the seed. Depending upon the cultivar, a kernel is either sweet or bitter.
9. Apricot kernel oil is an important commercial commodity which is quite similar to almond oil in physical and chemical properties both containing over 90 per cent of their fatty acids such as oleic and linolenic acid.

Area and Production

World- Area: 4.1 lakh ha, production: 27.5 lakh tones and productivity: 6.6 t/ha

India- Area: 13573 ha, production: 18267 lakh tones and productivity: 1.3t/ha.

Production is highest in turkey followed by Iran and Russian federation. In India production is highest in Himachal Pradesh followed by J&K and UP.

Taxonomy

Family: Rosaceae

Subfamily: prunoideae

Genus: Prunus

Subgenus: prunophora

Chromosome no: 2n=16

Most of the cultivated apricots belong to species *Prunus armeniaca*. Several species of apricot are classified either as sub species or as botanical varieties of *P. armeniaca*, but most of them are accepted as distinct species and are grown for fruits, rootstocks, oil or ornamental purpose.

Name of the Species	*Common Name*	*Where grown*	*Uses*
P. ansu	—	China and Japan	Fruit, rootstock
P. armeniaca	Apricot	Entire world	Fruit, rootstock and oil
P. brigantiaca	Briancon apricot	France	Fruit, oil
P. dasycarpa	Purple apricot	China,USSR	Cold and disease hardiness
P. mandschurica	Manchurian apricot	China, Korea	Cold hardiness
P. mume	Japanese apricot	China and Japan	Fruit, ornamental
P. sibirica	Siberian apricot	Siberia	Breeding purpose

P. dasycarpa is a natural hybrid between *P. armeniaca* and *P. cerasifera*

P. mume species of apricot come to bloom in the month of February and if the weather remains good the fruit set heavily. The fruits are however used in pickle, liquor making in Japan where it is mostly grown in warmer and humid conditions.

P. sibirica and *P. mandschurica* are very cold hardy species and with stand as low as 5°C during dormancy. Hence these are used for imparting the cold hardiness in the apricot breeding programmes.

Climate and Soil

Apricots can be successfully grown at an altitude between 900 and 2000 m level as the long cool winter (300 - 900 chilling hours below 7°C) and frost free spring are favorable for fruiting. The cv.Valenciano -3 needs 300 hours of chilling cv.bulida 1 and 2 needs 700 hours of chilling.

White fleshed sweet kernelled apricot require cooler climate and are grown in dry temperate region up to 3000 m above mean sea level.

Yellow fleshed bitter kernelled ones thrive better under the warmer climate (900-1500m).

Average summer temperature (16.6°C - 32.2°C) suitable for better growth and quality fruit production. Spring frost causes extensive damage to blossoms which are killed when temperature falls below 4°C.

The pH of soil should be 6.0-6.8. Being hardy, it can grow in most of soils. deep fertile and well drained loamy soils are more suitable for its growth and development.

Rootstocks

Apricot seedlings are mainly used as rootstocks the apricot and peach seedlings are suitable for sandy soils and dry conditions while plum rootstocks good for heavy situations.

Characteristics of Rootstocks

Below mentioned four groups are generally used for apricot

1. Apricot seedlings
2. Peach
3. Plum
4. Related species

Apricot Seedlings

Apricot rootstocks are resistance to root knot nematode and crown rot but it is not tolerant to poor soil drainage conditions.

Haggith is a seedling of apricot used as cold–hardy, vigorous and compatible with wide range scion cultivars.

INRA Manicot – a wild apricot selection as been assessed as a very uniform rootstock.

The seedling of North African wild apricot give good performance as a rootstocks in the soils having high salt and lime content.

Peach

Peaches are popular rootstocks for apricot mainly adapted to acidic to natural and sandy dry soils under irrigation.

Example –Important rootstocks are Okinawa and Nemared.

Plum

The plum rootstocks for apricot used mainly on heavier soils are under wet soils conditions.

Marianna plum cutting were found to be most tolerant to water logging than Myrobalan. Different clonal rootstocks have been recommended in different regions.

Examples - Myrobalan 29 C and Marianna 2624.

Related Species

1. The use of sand cherry (*P. besseyi*) as rootstock produces dwarf trees.
2. *P. sibirica* and *P. mandschurica* are used as rootstocks for colder region but *P. sibirica* did not reduce bud damage caused by temperature fluctuations in early spring.
3. *P. insititia* selected from a local population seedling was found vigorous and compatible with most apricots. It is resistant to low winter temperature.

Development of Seedling Rootstock

The seeds should be kept moist during the period of stratification. The germination of apricot seeds can be hastened by removal of kernel from shell, scarification, stratification and gibberellic acid or kinetin treatment.

1. For raising of rootstock, seeds are collected fully ripe fruits of wild apricots.
2. The germination of seeds can be hastened by soaking the seeds for 24 hrs in 500 ppm GA_3 before sowing.
3. Apricot seeds require stratification for a period of 45-50 days at 4°C to break dormancy. The stratified seeds are sown at a distance of 1520 cm from seed to seed in rows 25-30 cm apart.
4. After sowing the beds are mulched with 6-10 cm thick grass and light irrigation is applied. The seedlings attain a graftable size one year after sowing.

Propagation

1. Tongue grafting: The seedlings of pencil thickness are grafted with tongue method in February.
2. T-budding generally adapted for its multiplication in the month of June.
3. Chip budding performed in September also very good success.
4. After care of grafted plants like single stemming, staking, weeding, watering, and plant protection measures should be adapted at regular intervals.

Planting

1. Apricot is planted during the dormant season (December end to mid-March) but early planting gives better establishment of plants.

2. The spacing of plant varies with the soil climate and vigour of the cultivar to be grown. square planting system with a distance of 6 x 6 meter is traditionally adopted.
3. There is, however, a recent trend towards rectangular pattern because more plants get accommodated in row in comparison to traditional wide spacing (8 x 6 and 10 x 6 m) and a closer spacing of (7 x 5m) gave higher yield in apricot.

Training and Pruning

Training system such as modified central leader system is adopted for training of apricot plants.

Pruning of Young Plants

1. The main stem of one-year-old plant is headed back at about 70 cm above the ground level and only three to five well spaced side shoots are allowed to grow in all directions which later on turn into main scaffold branches.
2. The pruning becomes more important in the second year because the frame developed in this period, will give ultimate shape to the tree. The branches chosen to become scaffold branches are headed back at different heights and all other branches are removed from the base.
3. In the third year, the scaffold and lateral branches on them have much growth. Some shoots may also emerge directly from the trunk. The scaffold branches along with few laterals are allowed to grow by removing all other shoots. If the leader branch is making excessive growth it should be headed back to enhance the growth of other main limbs.
4. Thereafter, in the fourth year the pruning comprises the thinning of overcrowding and crisscross branches.

Pruning of Bearing Trees

1. The apricot bear on spurs and laterally on one-year-old shoots. The production off young growth is, therefore, essential for the initiation of new spurs which generally takes place at the base of growing laterals.
2. Thinning of some one-year-old shoots combined with one-third shortening of the other shoots had given the highest yield than the other pruning methods.
3. The shoots should make 50- 60 cm growth each year in young trees and 25 – 35 cm in older ones.
4. The old branches and diseased branches should also be removed to replace with younger ones.

Irrigation

The peak water use period is from April end to mid-June, which coincides with the fruit development period. In Himachal Pradesh, 3 irrigations in season are sufficient for size and quality fruits. Hay mulch (10-15 cm thick) and black plastic mulch also help conserve soil moisture.

Flowering

In apricot usually three buds develop in the axil of at each node on spur the central one being a vegetative bud the two side buds are floral. High temperature humidity and wind shorten the flowering duration by increasing the transpiration. Apricots suffer greater damage due to spring frost because of blooming earlier than any other temperate tree fruits.

Pollination and Fruit Set

Most of the commercial apricot cultivars have been found self-fruitful in pollination studies and set fruits without pollinizers but open pollination has always given better fruit set. The cultivars Charmagz, Perfection, Riland are self-incompatible. The provisions of 2-3 bee hives/ha in apricot orchard for good fruit set is required.

Growth Regulators to Increase Fruit Set

1. Gibberellic acid (GA_3) 10ppm when sprayed at blossoming stage increases fruit size.
2. A spray of 50 ppm 2,4,5-T P at the beginning of pit hardening also reduced post harvest drop.

Fruit Thinning

Fruit set in apricot is rather heavy which results in tendency of biennial bearing. Fruit show increasing tendency of biennial bearing. Fruit thinning improves fruit size promotes regular bearing, decreases limb breakage (due to heavy crop load) and maintains the tree vigour.

1. Fruit thinning should be done within 40 days after full bloom (last week of April or first week of May).
2. A spur should not have more than 2 fruits.
3. Foliar spraying of 50 ppm NAA at 20 days after fruit set is ideal for thinning.

Fruit Growth and Maturity

The apricot fruit have **double sigmoidal growth pattern** having a retarded growth period at the time of pit hardening. Fruits develop rapidly and ripen early in summer season. Temperature of about 23°C accelerates the developmental processes of the fruits.

Harvesting

Apricot fruits generally mature during first week of May to last week of June depending on variety and location.They are harvested manually and no mechanical harvesting is practiced.

Maturity Indices

1. Change in ground colour from green to yellow
2. Total soluble solids 10 per cent
3. Titrable acidity 0.4 to 1 per cent
4. Days to full bloom 100-120 days
5. Firmness 5-7 kg/cm^2

Since apricots are very perishable, due care is required during harvesting and packing and transportation. The fruits should be harvested in morning hours and precooling has to done at 5°C. Direct exposure of fruits to sun should be avoided during grading and packing.

Yield

Apricot trees start fruiting at the age of 5 years and continue up to 30-35 years They attain full bearing age at about 7-10 years, yielding 50 - 80 kg/tree or 15-22 t/ha.

Storage

Refrigerated Storage

Although apricots are perishable, they can be stored at 0°c for 1-2 weeks with 85-95 per cent relative humidity.

CA Storage

In CA storage Apricot can be stored at 2-3 per cent CO_2 and 2-3 per cent O_2 for 7 weeks at 0-5 per cent.

Chapter 7

Almond

Introduction

Cultivated Almond *Prunus dulci* or *Prunus amygdalus* is native to cool winters, frost free spring and warm dry summer. The Almond is one of the major and oldest tree-nut crops in the world.

The USA and Spain produce more than 50 per cent of the world production followed by Italy, China and Iran.

Origin

1. Hot arid region of West Asia.
2. According to a Russian scientist, cultivated almond is a hybrid between two species *Prunus fenziliana* × *Prunus bucharica*.
3. Wild species of Almond is *Prunus webbii*.

Composition and Uses

Almond Kernel is a Very High Energy Source

Protein	20.8 per cent
Fat	58.9 per cent
Mineral matter	2.9 per cent
Fiber	1.7 per cent
CHO	10.5 per cent

Ca	0.23 per cent
Iron	3.5 per cent
P	0.49 per cent
Vitamin B_1	240mg/100g

Kernel of Bitter Almond

Lipids	55.6 per cent
Fat	18.9 per cent
Crude fiber	2.0 per cent
Ash	2.3 per cent
TSS	7.9 per cent
Hydrocyanic acid	0.27 per cent

Uses

1. The nuts are usually eaten when they are ripe. Kernels obtained after blanching, roasting, frying and salting are highly delicious and in great demand.
2. Almond butter is free from starch which has good demand for diabetic patients.
3. Almond oil is used in confectionery and also for pharmaceutical and cosmetic preparations.
4. Almond butter free from starch has a good demand for diabetic food. However, green nuts can also be consumed.
5. Green nuts are picked before the outer husk begins to loose its colour and harden. The nuts are then opened and the kernels eaten in the milky stage with slightly sugar cream cheese.
6. Bitter Almond: Kernels are used in perfume, pharmaceuticals and a small amount is used for making pastries.
7. Kernels contain **Amygdalin**, a cyanogenic glucoside which differs from all the other known cyanogenic glucosides in having Gentiobiose as its sugar unit.
8. Very valuable oil called **Badam Roghan** is extracted from Almond and it is considered to have high medicinal value which is similar to Apricot oil.

Area and Production

In India, Area: 0.012 m.ha and Production: 0.08 mt (NHB Data, 2017-18).

Bitterness in almond is due to **single recessive gene**.

Botany

Almond (*Prunus amygdalus or Prunus dulcis*) belongs to the

Family: Rosaceae

Genus: Prunus

Subgenus: Amygdalus

Chromosome no. 8

Morphologically and taxonomically, Almond is very closely related to peach. The tree characters of peach and almond are very similar. The chief differences in the two are found in mature fruits.

1. The fruits of peach are delectable succulents, but of almond inedible.
2. The flesh of peach is fleshy, juicy and soft, that of almond thin, tough and leathery.
3. The pit of peach must be removed, while that of almond drops off naturally from flesh which splits at maturity.
4. The pit of peach is deeply sculptured and of bone-like consistency that of almond is nearly smooth and in some varieties thin and soft in texture.
5. Both almond and peach have low chilling requirement. Almond, in general, needs somewhat lesser chilling than peach as it sprouts earlier.

Climate

1. The limiting factor in almond cultivation is spring frost, particularly during full bloom, it is often responsible for an irregular yield.
2. Almonds can be grown successfully both in temperate and subtropical areas with low rainfall. Elevation of prominent almond-growing areas ranges from above 700 to 2500 meters above sea level.
3. The chilling requirement of almonds for proper growth and fruiting are rather low, *i.e.,* an exposure to low temperature below 7.2°C for 250-500 hrs during winter is only needed. It can tolerate temperature as low as -20°C but at 2°C damage may occur to the buds and flowers.

Soil

1. Soils with good drainage are essential, in waterlogged soils, even well grown almond tree collapses suddenly whenever the soil remains saturated for 3-4 days.
2. Light loamy soils are considered suitable for almond while heavy soils are not desirable. The ideal soil is the deep alluvial type.
3. Almond, can tolerate high summer temperatures and very low humidity better than any other temperate fruit tree. With its deep root system up to

3m and leathery leaves resistant to the drying up, almond can withstand long periods of summer dry weather.

4. It is important to irrigate almond particularly when it is grafted on peach rootstock.

Rootstocks

Almond Seedling

Almond seedling rootstock from the bitter or sweet cultivars are still used because of their tolerance to drought, lime soils and iron chlorosis, and their longevity, but their rootstocks are susceptible to several diseases especially in irrigated soils.

In winter climates and irrigated areas, peach seedlings are used. They have more uniform vigour and survival during transplanting is higher.

Selected Sootstocks

1. Peach × Almond hybrids (GF 677; GF 557)
2. Plum (Myrobalan 2032, Marianna 2624, Marianna GF 8/1) and
3. Hybrids of Myrobalan × Peach (or) Myrobalan × Almond.

Hybrids of Peach × Almond

1. GF 677 hybrid is almost exclusively used in intensive almond orchards.
2. It induces rapid growth and precocity but it is very susceptible to nematodes.

For the potential rootstocks only Marianna 2624 and Marianna GF 8/1 are used for intensive orchards

1. Both are used for poorly – drained soils subject to waterlogging.
2. Both show good resistance to nematodes and are easily propagated asexually.

Hybrids between *Prunus cerasifera* × *P. persica* and *P. cerasifera* × *P. dulcis* are difficult to obtain and they show a high degree of incompatibility with almond. Until now they have not been used commercially.

Propagation

1. Seed
2. Grafting on seedling of almond, peach or plum
3. Grafting on selected rootstock

Raising of Seedling

1. Almond nuts are harvested from July-September and stored in cool and dry places till December.
2. Stratification is needed to break the rest period (2-3°C for 2 to 3 months).
3. By the middle of February the kernels begin to germinate.
4. Germinated seedlings can be placed at a space of 60×10 cm.

Vegetative Propagation

1. When seedlings are of pencil thickness (at height of 30 cm) then T-budding is done in the month of June.
2. Tongue or cleft grafting can be done in December-January.

Planting

Seedlings or grafted plants on one - or more year-old with well formed wood, healthy roots are planted in well-prepared soil fertilized with organic fertilizer. The best period for planting is between November-December.

The distance between trees varies according to the variety, rootstocks, and soil fertility and irrigation system, 5 × 5 m seems to be the best distance for cultivars grafted on peach seedlings or on peach × almond hybrids.

Cultivars

1. Non-pareil: Best variety of almond.
2. Merced
3. Price
4. IXL
5. Ne Plus Ultra: Acts as pollinizer for Nonpareil.
6. California Paper Shell
7. Shalimar
8. Makhdoom
9. Waris
10. Pranyaj
11. Butte
12. Carmel
13. Drake: Acts as pollinizer for Non-pareil.
14. Monterey
15. Peerless

Training and Pruning

Pruning is carried out at three different stages:

1. During early growing
2. When cropping
3. To reinvigorate old plant.

1. The early pruning is practiced to give shape to the tree.
   a. It is carried out immediately after planting by cutting at 1 m from the ground, when the commonest 'vase' shape is required.
   b. The primary scaffold limbs (3 or 4) developing within 10-15 cm from the top will form the tree crotch.
   c. Summer pruning is carried out 2-3 times in order to retain shoots for main branches and to eliminate unwanted branches and suckers. Summer pruning should be carried out in the second year.
   d. Further, winter pruning will be required in the 3rd and 4th yrs when the tree should have 3-4 main branches each with 5-7 secondaries.
   e. To allow light penetration the centre of the tree should be kept free of shoots.
2. When the tree is cropping:

   Regular pruning should be carried out every year to ensure high yield.
   a. Almond flowers are usually located on spurs which live about 5 yr.
   b. Renew about one-fifth of fruiting wood every year to stimulate new shoots by pruning.
3. Reinvigorating pruning is carried out only when trees are old and unproductive.

The main branches can be cut back and many new shoots will grow out in the spring.

Manuring and Fertilization

In Kashmir for 10 yr old trees 450g of N, 650g of P and 600g of K is recommended.

Pollination

Although some cultivars such as Tuono, Genco, Ferrante, Palatina, Scorsa Verde and Mollisona are self-fertile; Almost all others are self-incompatible.

1. Usually pollinizers, possibly of commercial importance are used.
2. Pollination is mainly by honey bees, usually 1 or 2 colonies per ha, but 4 or 5 colonies will ensure good pollination.

Fertilization

A temperature of 15-23°C (Lower than that for other deciduous fruits), although it depends on the cultivar, is necessary for pollen germination and pollen tube growth, which reaches the embryo sac in 3 days.

Flower and Fruit Drop

Almond nuts drop at 3 distinct stages:

1. Immediately after fertilization.
2. Within 30 days after flowering.
3. After 45 days from flowering (in April).
4. Although flower and fruit drop always occur, improvement in management of orchards reduces the drop at different stages. Usually the percentage of harvested nuts is 15-35 per cent of initial number of flowers.

Fruit Growth and Maturity

1. The almond fruits are an egg-shaped drupe with a sigmoid growth pattern. Fruit development stops at 70-75 days after blooming.
2. The embryo grows very slowly at the beginning, when the fruit reaches full size the embryo grows very fast and uses all the endosperm which remained in a semi-solid consistency. At maturity the embryo, with its big cotyledons, fills the whole seed.
3. The pericarp (hull) is almost always thin (5-15mm) and may be pubescent. In July it begins to split open freeing the endocarp. The endocarp contains 1 or 2 kernels varying in size, weighing from 0.5 to 1.5g, with brown integuments at maturity.
4. Fruit thinning normally is not required.
5. The first sign of nut ripening is the dehiscence of the hull along the ventral suture and consequently the reduction of the removal force. Usually growers start to harvest one month and a half after the first sign of hull dehiscence.
6. Nuts growing at the periphery of the tree usually mature earlier than those located in the centre.

Harvesting and Yield

Almond can be harvested green or dry. Dry harvest is made from Aug-Oct. In non-irrigated orchards the nuts ripen earlier than in the irrigated ones.

Yield: 3-6kg nuts/plant

Yield of kernels: 54 per cent in Var. California's paper shell

46 per cent in Var. Pathick's wonder

Mechanical Harvesting

1. The orchards are often irrigated, 2 weeks before harvesting for better detachment of fruits.
2. Nuts are dropped by mechanically shaking the tree, possibly when the hull is still attached to the shell and swept into the rows from where they are picked up by machine and then conveyed to carts or trailers for hulling.

Storage, Grading and Kernel Use

After harvesting, the nuts are removed from the hull by machine.

1. The nuts are dried in the sun for 3-4 days until moisture content is reduced to 8.0-8.5 per cent, assuming that there will be a further 1 per cent loss during storage.
2. Storage of soft-shell nuts may become a problem due to the growth of mould.
3. Almond for table consumption are bleached with SO_2 to improve their appearance.
4. Shelling yield of cultivars varies according to the shell consistency

$$\text{Shelling yield} = \frac{\text{Kernel weight} \times 100}{\text{Kernel + shell weight}}$$

Shell consistency is generally classified as follows

- Stone or very hard shell (shelling yield = 20-25 per cent)
- Hard shell (shelling yield = 25-35 per cent)
- Semi-hard shell (shelling yield =35-45 per cent)
- Soft-shell (shelling yield =45-55 per cent)
- Paper-shell (shelling yield =55-65 per cent)

Grades or Kernels for the International Trade are

a. **Unselected kernels**: Consist of broken and double kernels containing impurities which after chopping or grinding, are used in chocolate and pastry industry.

b. **Selected kernels**: Selected according to the size and unit weight and after mechanical peeling, with water at 98°C for 2 min, are utilized for the food industry (confectionery, ice-cream, salted, sugar coated or toasted almonds).

c. **Broken, double, misshapen and bitter kernels**: Ground or in pieces are used in pastry, confectionery or for extracting oil used for cosmetics and in dermatology for its anti-inflammatory properties.

Virus and Virus like Diseases

1. **Strain of tomato ring spot virus:** It causes yellow bud mosaic.
2 **Strain of Prunus Necrotic Spot virus:** It induces Almond Calico.
3. **Infectious Bud failure**
4. **Non-infectious Bud failure**
5. **Non-productive syndrome**

Chapter 8

Chestnut

The chestnuts (*Castanea* spp.) have been an important item of human diet in northern hemisphere since time immemorial.

It belongs to family Fabaceae. It has been grown in China and Japan for more than 4000 years for food, fuel, timber and beautification.

The popularity of chestnut is attributed to its productivity, ease of harvesting, simplicity of preparation for eating, low fat content and diversity of high low income people.

Origin and Distribution

The different species of chestnuts are native to western Asia, Europe and Northern America.

Chinese chestnut (*Castanea mollissima*) is native to china.

Japanese chestnut (*C. crenata*) to Japan and South Korea.

Europe chestnut (*C. sativa*) is native to Mediterranean countries of south Europe and Asia Minor.

Western America is the native place for American chestnut (*C. dentata*).

Among all the major nuts, only chestnuts have a very low fat content and they have a higher proportion of pyridoxine.

Chinese chestnuts can withstand temperature upto -32°C.

Chestnuts are self-sterile and exhibits vivipary.

Common and Botanical Name, Chromosome Number and Range of Chestnut (Castanea) Species

Common Name	*Botanical Name*	*Chromosome Number*	*Origin*
Chinese chestnut	*C. mollissima*	24	China
Japanese chestnut	*C. crenata*	24	Japan, Korea
European chestnut	*C. sativa*	24	South Europe, Asia minor
American chestnut	*C. dentata*	24	East USA
Seguin chestnut	*C. seguina* *C. davidii*	24	China

Chestnuts have a blur like involucres that is green when immature and turn brown as the fruit matures. The mature involucre splits and releases the 1 to 4 enclosed fruits of the chestnut. The chestnut is ¾ to 1 ½ inches in diameter and rich chestnut brown in colour. Chestnuts can be eaten by slicing the shell and cooking them in microwave until the shell easily peels off.

Cultivars

Abundance, Nanking, Meiling, Orrin, Crane, Sleeping Giant, Hemming, Willosca, Jersey Gem and Marron de saint-vincent, Eaton, Ginyose, Colossal, Lyeroka, Maraval, Marigoule, Myoka, Olie, Orrin, Qing and Regina Montis.

Climate

Chinese chestnuts trees are cold-hardy than Japanese chestnut. Chinese chestnut can withstand temperature up to -32°C without injury when fully dormant. The annual rainfall in chestnut producing areas is 75cm.

Soil

Generally, chestnut is grown on wide range of soils, but the well drained deep and fertile soils are considered the best. Soil should be slightly acidic, with pH 5.6-6.5 chestnut does not thrive in water logged soils.

Rootstocks

The seedling of the same species should be used as rootstocks in chestnut. Chinese chestnut seedlings are used as rootstock for most of the cultivars due to the fact that these are resistant to chestnut blight and cold injury.

Seed Propagation

Chestnuts are mostly propagated through seeds, but the established cultivars are raised by vegetative method.

Vegetative Propagation

Budding, grafting, stooling, cuttings, top working and micropropogation

Planting

The best time for planting the chest nut trees is November-December.

Spacing: 12m ×12m

Fertilizer Recommendation

Application of N, P and K at 4.6 kg ammonium sulphate, 3.1 kg super phosphate and 0.8 kg M.O.P per hectare resulted in good growth and higher yields.

Flowering

Chestnuts are monoecious having pistillate and staminate flowers at different locations on the same trees. Staminate flowers appear on cylindrical catkins with 10-20 stamens and 6-parted calyx. Both types of flowers are borne in catkins on current seasons shoots.

Pollination

Since chestnuts are self-sterile mostly, more than one cultivar or seedling are planted together to ensure cross pollination.

Pollination occurs through wind and insects.

Duodichogamy

In chestnuts, the phenomenon of Duodichogamy is seen (first protandry followed by protogyny). A condition when two batches of male flowers are temporarily separated by a batch of female flower in between. The resting period in between avoids selfing completely. The unisexual staminate catkin occurs at lower parts of the shoot and the bisexual catkin at terminal end of the shoots. The lower unisexual staminate catkins bloom first, and then pistillates of the bisexual catkins bloom. This sequence of maturation of staminate, pistillate and then staminate catkin is known as **Duodichogamy**.

Fruit Development

The kernel consists of two cotyledons joined with a growing point and enclosed in a pellicle which is a membranous covering originating as integument of the ovule. Fruits ripen in early September in some cultivars and the latest ripening period is late October or November.

Harvesting

Since chestnuts are highly perishable, the nuts fallen from involucres to the ground are gathered every alternate day and the gathering period lasts from 16 to 30 days.

Yield: A 16 years old tree produces about 8.1kg of nuts.

Storage

The moisture level of the nuts is brought to 10 per cent by drying at 4.4°C with well circulated air of 70 per cent relative humidity.

Diseases

1. **Chestnut blight:** *Endothia parasitica*

   Chinese chestnuts are resistant to blight.

Chapter 9

Strawberry

Introduction

The strawberry is known as the most delicious and refreshing fruit of the world to millions of peoples. Its plant is also cherished in gardens and commercial plantations for its beautiful and attractive red fruits. Besides, its fruit is a rich source of vitamins and minerals with delicate flavour.

Nutritive Value and Uses

The fresh, ripe fruit of strawberry is a rich source of vitamins and minerals. It is fairly a good source of vitamin A (60 IU/100 g of edible portion) and vitamin C (30-120 mg/100 g of edible portion). Strawberry is also a rich source of pectin (0.55 per cent) which is available in the form of calcium pectate and serves as an excellent ingredient for jelly making. Fructose and glucose are major sugars in its fruit with a small portion of sucrose. Citric acid is the most abundant organic acid in strawberry, followed by malic, succinic and oxalic acids. **The characteristic aroma in strawberry is attributed to ethyl hexaonoate, methyl hexaonoate, ethyl heptanoate, ethyl propionate, ethyl butanoate, methyl butanoate, furanone and linalool.** Linalool and nonanal are major constituents of strawberry oil. The red colour of fruit is mainly due to the presence of anthocyanin, pelargonidin – 3-monoglucoside and traces of cyanidin.

Composition and Nutritive Value of Strawberry Fruit

Water (per cent)	85 – 90
Protein (g)	0.25 – 0.7
Carbohydrates (g)	8.5 – 9.2

Fats (g)	0.2 – 0.5
Fibre (per cent)	1.1
Fructose (per cent)	1.5 – 3.5
Glucose (per cent)	1.5 – 3.0
Sucrose (per cent)	0.8 – 2.5
Citric acid (mg/100 g of edible portion)	4000 – 1250
Malic acid (mg/100 g of edible portion)	90 – 675
Succinic acid (mg/100 g of edible portion)	90 – 100
Oxalic acid (mg/100 g of edible portion)	20 -24
Tartaric acid (mg/100 g of edible portion)	15 – 17
Vitamin A	60.0 – 63.0 IU
Ascorbic acid	30 – 120 mg
Thiamine	0.03 mg
Riboflavin	0.07 mg
Niacin	0.06 mg
Anthocyanins (mg/100 g of edible portion)	50 – 100 mg

Origin, History and Distribution

Origin of Modern Strawberry

The modern cultivated strawberry (*Fragaria ananassa*) is a hybrid between two largely dioecious octoploid species, *Fragaria chiloensis* and *Fragaria virginiana.*

History

The history of strawberry goes back to the Romans and perhaps even the Greeks, but because the fruit has never been a staple of Agriculture, it is difficult to find out an ancient reference to it. Theophrastus, Hippocrates, Dioscorides and Galen did not even mention it, nor did Cato, Varro, Columella or Palladius, the four world famous Latin writers on Agriculture. Apulius cited the strawberry only for its medicinal value. Although Virgil and Ovid did name the strawberry in their verses but they did so only casually in poems of the country life, where they associated it with other wild fruits. However, it is evident from literature that by 1300's, strawberry was in cultivation in Europe and then the French began to transplant the wood strawberry (*Fragaria vesca*) from the wilderness to the garden. But the plant was considered more ornamental for its flowers than the use for its fruits, although it was grown to some extent for table purpose also. There are many theories in England about the derivation of the name Strawberry. Anglo-Saxons called the strawberry as "hayberry" because it ripens at the time the hay is mown.

The 15th century has horticultural significance for establishing strawberry as a common garden plant in France. The common garden strawberry in France was *Fragaria vesca*, the wood strawberry. The other species of Fragaria like *F. moschata* (the musky flavoured strawberry), *F. viridis* (the green strawberry), F. *alba* and *F. sylvestris* were also described by Jerome Bock in 1532.

Taxonomy

Fragaria species belongs to family Rosaceae, with basic chromosome number of x = 7. The cultivated strawberry, *Fragaria ananassa*, is an octoploid, having chromosome number (2n) of 56.

Distribution of Strawberry Species and their Ploidy Level

Ploidy Level and Chromosome Number	*Major Species*	*Distribution*
2x = 14	*F. vesca, F. viridis, F. nilgerrensis, F. daltoniana, F. nubicola, F. iinumae, F. yesoensis, F. nipponica, F. mandschurica*	Central Asia and Far East
4x = 28	*F.moupinensis, F.orientalis, F. corymbosa*	East and South East Asia
6x = 48	*F. moschata*	Europe, North and South America
8x = 56	*F. chiloensis*	Central and North America, Chile, Peru
	F. virginiana	
	F. iturupensis	
	F. ananassa	

Species of Genus *Fragaria*

Species

1. Diploid (2n = 2x = 14)

Fragaria vesca (wood strawberry), is the most extensively distributed of all the species of *Fragaria*. Flowers are bisexual and generally self – fruitful.

The *Fragaria viridis*, a native of Europe, is found along the edges of forests. The leaves are deep green. It bears small inflorescence, and bisexual flowers, larger than that of *F. vesca*.

The *Fragaria nilgerrensis* is a native to south-east Asia and China. Its plants are robust and spreading, producing strong runners.

The *Fragaria daltoniana* occurs in a small area of the Sikkim Himalayas (India) at an altitude of 3,000 – 4,500 m.

The *Fragaria nubicola* is found in the then USSR and Sikkim Himalayas (India) at an altitude of 1,500 – 4,000 m above mean sea level. The plants resemble closely those of *Fragaria vesca*.

The *Fragaria iinumae* is found in the Alpine mountains of central and northern Japan.

The *Fragaria yesoensis* is native to Japan.

The *Fragaria nipponica* and *F. mandschurica*.

2. Tetraploids (2n = 4x =28)

The *Fragaria moupinensis* is one very important tetraploid species.

The *Fragaria orientalis* is chiefly found in western Siberia, Mongolia, Manchuria and Korea. The *Fragaria corymbosa* is probably a native to northern China.

3. Hexaploids (2n = 6x =42)

The *Fragaria moschata* is the only hexaploid species of *Fragaria* which occurs in northern and central Europe, Russia and Siberia.

4. Octoploids (2n = 8x = 56)

The *Fragaria chiloensis* is the most important octoploid species of *Fragaria*, found along the Pacific coast of Alaska, central California, Chile and Hawaii. The fruits are very large, dull to bright red firm, with white flesh and pungent flavour.

The *Fragaria virginiana* is another octoploid species of *Fragaria*, found throughout the central and eastern North America.

The *Fragaria iturupensis* is another octoploid species of strawberry.

Important Cultivars

The main characteristics of some of the strawberry cultivars are given hereunder:

1. **Belrubi:** Originated in France. Fruits are large, juicy with moderate flavour.
2. **Blomidon:** It was developed in Canada, as a result of cross between K72-4 and Micmac. Calyx is moderately difficult to remove. It has tart flavour. It is moderately resistant to powdery mildew, *Verticillium* wilt and resistant to *Botrytis* fruit-rot and red stele.
3. **Camarosa:** Plants are short-day type, vigorous, producing fruit over an extended period in arid subtropical climates.
4. **Calypso:** It is a day-neutral variety.
5. **Cambridge Favourite:** It is a cross between Avant Tout and Blackmore. The fruits are large, rounded and plump, pinkish-scarlet, turning light red on maturity.
6. **Cambridge Prizewinner:** A cross between Early Cambridge and Howard 17, Cambridge Prizewinner is an early maturing variety bearing attractive red fruits.
7. **Chandler:** It is highly acceptable present day cultivar of northern plains of India. This is a high yielding short day variety.
8. **Fern:** It is an early day-neutral cultivar developed in California.

9. **Huxley:** It is a mid to late season cultivar.
10. **Pajaro:** Pajaro, a seedling of Cal 63.7-101 × Sequoia developed in California.
11. **Red Rich:** It is also known as Red Glow and Hagerstorm's everbearing strawberry.
12. **Selva:** It is a day-neutral variety, which produces fruits both in open, under low plastic cloches or in peat bags under greenhouses.
13. **Senga Sengana:** It is a seedling selection from a cross between Markee and Sieger.
14. **Sweet Charlie:** It is a cross between FL 80 – 456 and Pajaro, which was released for commercial cultivation in Florida during 1997.
15. **Tioga:** A cross between Fresno and Torrey. Tioga has short harvest season. It is also difficult to pick for the fresh market because the fruit tends to 'cap' easily and come off without calyx. **It is suitable for processing but not for fresh market.**
16. **Torrey:** A sister seedling of Tioga, the plants of Torrey are of medium vigour, producing red fruits.
17. **Tribute:** This is perhaps the first day-neutral (everbearing) cultivar developed in USA.
18. **Tristar:** It is a sister seedling of Tribute. It is also a day-neutral cultivar.
19. **Vantage:** A cross between Tioga and Vestar. It is a high yielding cultivar, the fruits of which are suitable for fresh and processing purposes. The plants are resistant to *Botrytis cinerea* and highly resistant to *Verticillium* wilt.

Soil and Climate

Soil

Strawberry can be grown on a variety of soils, ranging from heavy clay to gravel soils, but it prefers light porous soils, which are rich in humus. It prefers slightly acidic soils with a pH of 5.8 – 6.5. On alkaline soils, strawberry may not establish properly and it may result in leaf chlorosis due to lack of iron absorption by the roots.

Climate

a. Light

Flowering in strawberry is strongly influenced by photoperiod, temperature and photoperiod × temperature interaction. On the basis of photoperiodic requirements, strawberry cultivars have been categorized as short-day, long-day or day-neutrals. In short-day cultivars, floral induction occurs with photoperiods of less than 14 hours, though these cultivars flower continuously regardless of day length but the environmental temperature should be less than 16°C. The day-neutral cultivars, oftenly called as everbearer, flower continuously, regardless of day length. Generally, flowering in strawberry occurs under the following photoperiodic regimes, short light period (10 hr) with long dark period (14 hr), short light and dark periods (10 hr), and long light and dark periods (14 hr). Present-day strawberry is grown extensively in temperate and to some extent in subtropical culture, *i.e.* short-days in autumn and hard winter or to Meridional culture (long days in autumn and moderate winter). The cultivars like Gorella, Redgaunlet and Senga Sengana grow under Septentrional conditions, while Tioga and Torrey are important cultivars of Meridional climate. However, with the development of several day-neutral cultivars like Selva, Fern, Tribute, Tristar, *etc.*, it is possible to grow strawberry throughout the year, but only in temperate climate.

b. Temperature

An optimum growing temperature of 15°C has been reported for most of the strawberry cultivars and species, though it grows well at a temperature range between 20°C and 26°C. Cold injury to open flowers often occurs at 2°C, although flowers of several cultivars survive even at -4.5°C.

c. CO_2 Concentration

The concentration of CO_2 plays a vital role in strawberry production under protected cultivation. Exposure of the plants to elevated CO_2 concentration typically enhances vegetative growth and yield, although concentration above 1000 (µM) may result in stomatal closure and decreased vegetative growth. In general, CO_2 concentration of 750 (µM) is optimum for most of strawberry cultivars grown under greenhouse conditions.

Propagation

Strawberry can be propagated through sexual (seed) and asexual (vegetative means). Propagation by seed is not considered as a viable method of propagation for cultivated strawberry as the plants raised through seeds do not come true-to-type. However, strawberry is usually propagated through runners. Nowadays, its large-scale propagation by tissue culture is being used widely.

Seed Propagation

Like many other fruit crops, propagation of strawberry through seeds is not considered as a right method of propagation. However, hybrids are first raised through seeds only. A single berry produces thousands of seeds, so a strawberry breeder has an advantage over other fruit crops in the sense that a number of hybrid seedlings can be produced with a single attempt. Seeds are very small and do not germinate properly unless they are stratified for a certain period at a given temperature. The effective chilling temperature (the temperature required for breaking rest) ranges from -2°C to 6.5°C but a chilling temperature of 9.5°C to 10°C is apparently most effective. The duration of chilling period is usually 2-4 weeks. Further, the treatment of the seeds with H_2SO_4 or GA_3 and Thiourea also breaks the seed dormancy and facilitates better seed germination.

Vegetative Propagation

The stolon, a creeping stalk, is produced in the leaf axil and grows out from the parent plant during summer. At the second node, a runner plant is formed and a new stolon arises on the runner plant to continue the runner train, which often branches. Initially, the runner plant produces a few roots and thereafter makes excessive fibrous roots. Thus, the strawberry naturally propagates itself by vegetative method of runner production. After the runner plant has acquired sufficient growth and roots, it may be separated from the mother plant and can be planted elsewhere. Although, propagation by runners perpetuates all the characters of the mother plant, viral diseases are quite often transmitted through runners. Thus, for runner production, a separate bed should be used. The site and soil where strawberry had not been grown for the last 3-4 years should be selected for the runner production. The planting for runner production should be done at 1.2 m × 1.2 m or at 1.8 m × 1.8 m distance.

Planting

Planting can be done either on furrows or on raised beds in a particular system of planting. In India, raised beds are usually preferred.

Planting Systems

1. Hill System

This system is commonly followed in cultivars developing few runners. In this system, the plants are grown either on single or double rows on the raised beds. The beds are usually made 15-20 cm high and runners are set 20-25 cm apart in twin rows, 30-35 cm apart and a distance of 75-100 cm is kept between the rows. This system demands a much higher economic input and a large number of runners are set per unit area and thus makes removing of runners difficult.

2. Spaced Row System

This system of planting is used for varieties which are moderate to weak in producing daughter plants. In this system, runners are usually set 30-50 cm apart in rows with spacing of 90-100 cm between the rows. This system of planting helps in easy picking of the fruits by moving between the rows easily.

3. Matted Row System

This system is adopted in areas where crown injury occurs in winter either due to freezing or temperature fluctuations. The runners are planted along the rows about 90 cm apart with 45 cm space between the plants. The width of mats is maintained between 40 and 45 cm. In this system, more number of plants can be accommodated per unit area. This system is suitable for obtaining heavy crop of medium sized fruits for processing purpose.

Time of Planting

1. It can be done at any time from July to following April or May. Early planting ensures a good first year crop.
2. Planting during July-August needs adequate care particularly from strong sunshine or dry soil conditions.

Irrigation

1. Strawberry is a shallow rooted plant and the roots are found within 15-25 cm of the soil surface.
2. Irrigation at pre-bloom and subsequent flower bud formation is critical.

Mulching

1. Both organic and inorganic mulches are used for strawberry cultivation.
2. Mulching protects fruits from soil contact and reduces the number of dirty fruits and prevents the occurrence of fruit rot.

Flowering

The flowers of strawberry occur in cluster. Each flower contains five sepals and five petals. The inflorescence is raceme and has a dichotomous branching. A typical inflorescence bears primary, secondary, tertiary and quaternary flowers. Two types of flowers, *i.e.,* hermaphrodite and pistillate flowers occur, in the cultivated strawberry. On the other hand, three types of flowers, *i.e.,* staminate, pistillate and hermaphrodite flowers are found in wild hexaploid and octoploid species.

The photoperiodic response of strawberry is well recognized. In general, the flower bud initiation takes place under short-day conditions. Runners and increased leaf size favoured under long-days. Purner *et al.* (1984) studied the effects of photoperiods and temperature on strawberries. Based on photoperiod they are classified as:

1. **June Bearers** (Spring crop strawberries): They produce flower buds when the days are short and the nights are long. They set flower buds in the autumn, during the short day period. These are generally used for processing, *e.g.,* Red chief and Guardian.
2. **Ever bearing varieties:** They are usually long-day and short night sensitive. They produce flower buds while the days are long and quit producing flower buds when the day length shortens in the fall, *e.g.,* Owrown, Ozark Beauty, Geneva and Superfection.

3. **Day-Neutral varieties:** Not sensitive to day length having 3 peaks of production, *i.e.,* early summer, midsummer, and early autumn, *e.g.,* Selva, Fern, Tristar and Tioga, *etc.,*

Pollination

In strawberry, ovule fertility and receptivity exist for 8 to 10 days, even for longer time in good weather. Anthers dehisce gradually over several days releasing pollens on warm day, mostly afternoon.

1. Although achene fertilization is possible through self-pollination, cross pollination by insects is useful in strawberry since stigmas become receptive before the anthers release pollen.
2. Incomplete pollination may result in developing irregular and poorly shaped berry.

Fruit Set

1. After fertilization, the receptacle swells and become fleshy and turns into a berry. The achenes are dry, small, hard and indehiscent.
2. Maximum number of well shaped fruits developed from tertiary flowers than primary or secondary flowers.
3. Strawberry is an aggregate fruit formed by the ripening of several ovaries belonging to single flower and adhering as a unit on a common receptacle.
4. Strawberry is a non-climacteric fruit.

Harvesting and Yield

1. Strawberries are generally harvested when fruit develops full colour.
2. Delaying in picking usually increases the proportion of overripe and rotted berries.
3. Berries should be picked along with a small stem portion attached.
4. Picking should be done in the morning as it facilitates better shelf-life.

Pick–your Own Berries (PYO)

In most of the advanced countries, labour is always in short supply. Thus, shortage of labour and competition from distant areas has forced many growers to adopt the pick-your-own method of harvesting.

Tioga – 96.53 quintals per hectare

Torrey – 47.52 quintals per hectare

Physiological Disorders

1. Albinism

a. It is the most serious disorder of strawberry fruits, occurring primarily at the time of ripening.

b. Fruits suffering from albinism appear bloated and develop white or pink areas on their surface and the pulp remains pale.

c. The fruits have poor flavour and tend to be more acidic.

d. The affected fruits develop normally but not ripen uniformly and show waxy appearance.

e. These fruits are liable to severe damage during harvesting and become highly susceptible to fruit-rot during storage.

f. Albinism tends to develop in densely growing crops, dense plantings or due to excessive fertilizer application.

g. It generally develops in crops grown in sandy, low pH soils and soils with high N, K Ca contents.

h. Selection of a suitable variety, avoiding dense planting and excessive application of fertilizers, particularly N, K, *etc.*, are some measures to reduce its menace.

2. Fasciation

a. The abnormal flattening and enlargement of stems of fruits and witch's broom appearance of the plants is a major trouble in strawberry.

b. The affected plant produces little marketable fruits and no runners.

c. It results from unfavourable growth conditions in late fall, when days become too short for normal development of a particular variety.

d. Fasciation is a varietal characteristic as some varieties like Missionary and Black more never fascinate.

e. Fasciation may also result from insect damage, disease attack and lack of pollination, humidity and frost.

3. Fruit Malformation

a. Production of malformed fruits can be commonly seen even in recognized varieties of strawberry, in which common shape of berry is changed.

b. Usually, primary and secondary flowers produce more malformed fruits due to the undeveloped achenes at the distal end of the receptacle than the tertiary and quaternary ones.

c. Planting more vigorous plants, high nitrogen levels, insufficient pollination and lack of growth promoting substances are the major ones.

d. Provision of adequate pollinizer and honeybee hives in the commercial plantings may reduce the problem of fruit malformation.

4. Phyllody

a. Phyllody is the main abnormality of flowers and fruits of strawberry.
b. It is mainly caused by mycoplasmal infection.
c. In this malady, the flowers and fruits get flattened and fascinated.
d. It first affects the opening of flowers and subsequently fruit set.
e. Since, phyllody is difficult to control, it is better to uproot and burn the infected plants

Chapter 10

Macadamia

Macadamia (*Macadamia integrifolia*) is one among the limited number of well-known tropical nuts of the world. It is delicious and nutritious nut indigenous to eastern Australia. It is also called **Australian nut or Queensland nut** and is popularly grown in Australia, Hawaii, California, Florida and South Africa. Cultivation of macadamia has been attempted with good success in some isolated orchards in Kerala, Tamil Nadu, Karnataka and Odisha.

Macadamia integrifolia (smooth) and *M. tetraphylla* (Rough) are cultivated but *M. integrifolia* is more common and known as smooth shelled type which has almost spherical, smooth surface kernels while *M. tetraphylla* is known as rough shelled type. The macadamia nuts resembles cashew in taste but is more oily and globular in shape. The kernels contain relatively large amount of proteins (about 10 per cent) and oil (about 75 per cent).

Macadamia nuts are considered a good source of calcium, phosphorus and iron.

Family: Proteaceae.

Basic chromosome Number x =14, 2n = 28.

Origin: Rain forests of south-eastern Queensland and North-eastern New South Wales in Australia.

Major producing countries are USA, Australia, and South Africa.

Climate and Soil

Frost-free regions receiving an annual rainfall of about 120cm are well-suited for growing this nut. Experimental evidences shows that macadamia puts up optimum tree growth at 25°C. pH of 5.5-7.5 may be preferred for the macadamia cultivation.

Varieties

***Macadamia integrifolia*:** Kakea, Ikaika, Keaau, Keauhou, Kau, Purvis, Makai, Mauka.

***M. tetraphylla*:** Greber, Renown, Anamour, Mammoth, Sewell and Probert 2

Hybrids: Beaumont, Nelinak 1 and Nelmak 2

It is better to plant at least at 2 cultivars in an orchard, although self-pollination occur. Kakea and keaau grow well under Bangalore conditions.

Propagation

Macadamia is propagated by wedge grafting on seed rootstocks.

Planting

Best season is autumn or spring. Trees should be spaced 9-10 m apart.

Training and Pruning

Preferable training method is strong central leader system.

Manuring and Fertilization

An annual dose of 450:150:500g of N: P_2O_5:K_2O/tree (in 2 splits) is recommended in addition to 40-50 kg of farm yard manure

Flowering

Flowers are born on long pendulous racemes. Protandrous and partly self-incompatible. Pollination through honeybees.

Harvesting and Postharvest Management

On an average, 7-year-old trees stars bearing. Under Bangalore condition, flowering occurs during December-January and fruits are ready in July-August.

The fruits are borne in clusters and on maturity. The husk begins to dry and splits exposing the inner brownish nut. The naturally drop and can be gathered by hand.

Yield

In Hawaii, a yield of 45kg nuts/tree after 18 years of planting has been recorded. Best managed orchard may give 3.2-3.5 t/ha.

Storage

Once the fruits are harvested, the husking and drying operations should begin immediately. After cracking the hard shell, raw kernels are dried to about 1.5 per cent moisture, which can be held satisfactorily for about a year.

Chapter 11

Cherries

Cherries occupy an important position among temperate fruits all over the world. The cultivated cherries are divided into two main groups **are sweet cherries (*Prunus avium*) and sour cherries (*Prunus cerasus*). The sweet cherry is mainly used for table purpose and sour cherry for processing.**

Taxonomy

Family: Rosaceae

Genus: Prunus

Subgenus: cerasus

Basic chromosome number: x=8

Composition and Uses

Cherries are rich in protein, sugars. It has more calories than apple. The carotene and folic acid content are also high. The fruit is a rich source of potassium, calcium, magnesium, iron and zinc. **Methyl anthranilate and Methyl salicylate** are the flavouring ingredients of cherry juice and the colouring principle of the fruit skin is **Keracyanin chloride** which is a diglucoside of cyanidin.

1. The sweet cherry is mainly consumed as dessert and also in canned fruit cocktails confectionary, ice-cream, bakery, juice making and for other various purposes.
2. The sour cherry is used for cooking, juice making, fruit salads, and distilling liquor.
3. Since the fruits are very perishable, these are largely preserved by canning, freezing and sun drying.

4. Cherries are brined with 1.5 per cent sulphur dioxide and 0.9 per cent calcium carbonate or hydroxide. Both fresh and frozen fruits are utilized for making cherry.

Origin and Distribution

1. Three species *viz.*, *Prunus avium* (sweet cherry), *P. fruticosa* (ground cherry) and *P. cerasus* L. (sour cherry) are involved in the development of present-day cherry form.
2. Cherries are **considered native of south central Asia.**
3. *P. avium* is usually diploid (2n=16) but triploid and tetraploid (2n=24 and 32 forms are also found).
4. Tetraploid Hybrids (2n=4x=32) between sour cherry and sweet cherry have been known as **duke cherries (*P. gondouini*).**
5. Tree and fruit characteristics of duke cherries are intermediate of sweet and sour cherries.
6. The ground cherry might be probable parent species of both sweet and sour cherry.

Area and Production

The leading cherry producing countries are USSR, USA, West Germany, Italy and France. In India cherries are mainly grown in the state of Jammu and Kashmir, where its total production is 605 metric tons from an area of 1110 hectares.

Climate, Soil and Rainfall

It requires 1000-1500 hours chilling period. During winter cherry blossom is very sensitive to spring frost.

1. An annual rainfall of 100-120 cm well-distributed throughout year is desirable but high rainfall during flowering results in heavy blossom wilt.
2. At the time of fruit ripening heavy rains causes fruit cracking. Therefore it should be dry at the time of fruit ripening.
3. Well-drained deep sandy loams with pH 6.5-7.0 can hold moisture during summer are most suitable.
4. The cherry plant is very sensitive to waterlogging. So heavy soil should be avoided.

Sweet Cherry Varieties

1. Early rivers
2. Early purple black heart
3. Emperor francis
4. Royal Ann (Napoleon)

5. Compact Lambert: It is a radiation induced mutant of Lambert.
6. Stella: First self-fruitful sweet cherry cultivar.
7. Compact Stella : It is a radiation induced mutant of Stella.
8. Van
9. Hedelfinger
10. Rainier

Sour Cherry Varieties

1. English Morello
2. **Montmorency:** It is a commercial variety
3. Meteor
4. North star

Propagation and Rootstocks

1. Locally known **Paja (*P. cerasoides*),** commonly used as a rootstock for sweet cherry in Himachal Pradesh, Kashmir and hills of Uttar Pradesh.
2. In some areas of Uttar Pradesh wild bird cherry (*P. padus*) is also used as a rootstock. Seedlings of mahaleb (*P. mahaleb*) are commonly used as a rootstock in Jammu and Kashmir.
3. Clonal rootstock that is Colt (*P. avium* × *P. pseudocerasus*) and Mazzard F12/1- have been found promising. They are recommended for raising dwarf sweet cherry plants in Himachal Pradesh.

Vegetative Propagation

1. In cherries vegetative propagation is mainly done by stooling, suckers, air layering, cuttings, budding, grafting and micropropagation.

Planting

A spacing of 6m x 6m is recommended for plants raised on seedling rootstock. Planting distance of 4m x 4m is recommended for plants raised on colt rootstocks in Himachal Pradesh. December-January is the planting time.

Pollination

Since most cherries varieties are self-sterile they need cross pollination. The universal pollen donor varieties are stella, vista and vega can be planted with any variety to get good fruit set provided their flowering period overlaps.

Training

Cherry trees are trained on **modified leader system.**

Pruning

Cherry plants require more corrective pruning rather than too much heading back of the branches. Fruits are borne laterally on spurs of **one-year old shoot**.

Manures and Fertilizers

Age of Tree	*Farmyard Manure (kg)*	*Calcium Ammonium Nitrate (g)*	*Super-phosphate (g)*	*Muriate of Potash (g)*
1	10	200	160	100
2	15	400	320	200
3	20	600	480	300
4	25	800	640	400
5	30	1000	800	500
6	35	1200	960	600
7	40	1400	1120	700
8	45	1600	1280	800
9	50	1800	1440	900
10 and above	60	2000	1600	1000

Farmyard manure should be applied in December along with a full dose of superphosphate and muriate of potash. Half dose of N is applied in spring before flowering and the other half one month later. Fertilizers are broadcasted in tree basin 30 cm away from them.

Harvesting and Postharvest Management

The fruit growth follows a double sigmoidal growth curve. Fresh fruits are picked with stem when the surface colour changes from green to red. While for processing, fruits are picked without stem.

Maturity Indices

1. Colour

The colour development is the best guide. Though it varies with the cultivar and the purpose for which the fruit will be used.

2. Total Soluble Solids

1.1 – 24.4 per cent

3. Acid Content

During ripening of sweet cherries, gradual reduction of acid content takes place.

4. Increase in Weight

Cherries gain in weight from the time they begin to mature until the fruit was fully ripe. Cherries continue to gain weight even after attaining full maturity.

5. Dessert Quality

Fruits of early picking varieties do not taste good and those of late picked varieties are very sweet or insipid. Those picked at right stage have better size, colour and appearance.

Chapter 12

Persimmon

Scientific Name: *Diospyros kaki*

Family: Ebenaceae

Chromosome No; X=15 somatic number = 60, 90

Origin: China

Fruit

A large juicy berry 1-10 seeded enlarged calyx at the base. Seeds large and flattened. Persimmon is considered as national fruit of Japan. In India, persimmon was introduced by the European settlers somewhere in 1921. At present, it is being grown on limited scale in J&K, HP, Hilly areas of UP & eastern part of eastern India.

Composition

Persimmons are high in vitamins A and C containing 2710 IU of vitamin A.

The characteristic astringency of immature persimmons is due to water soluble tannins (kaki tannin) present in the flesh of fruit.

Uses

1. The dried kaki is a traditional feature of New Year festival season in Japan.
2. In larger commercial operations, the fruits are first fumigated with sulphur dioxide and then artificially dried at 32-33°C.
3. Green fruits from strongly astringent cultivars such as Tsurunoko and Tsunagaki are used in the production of tannin of high quality.

Cultivars

Classified into two major groups

1. Astringent
2. Non-astringent

In astringent types water soluble tannins which cause astringency in the flesh, decrease as the fruits softens and becomes edible. Astringency can also be removed by various chemical treatments but these cause variable fruit ripening.

Astringent Cultivars

Hiratanenashi, Hachiya, Aizumishirazu, Yotsumizo and Yokono.

Non-astringent Cultivars

Fuyu, Jiro, Gosho and Suruga.

Climate and soil

1. Well drained sandy loamy soils are best.
2. Trees are deciduous and enter a rest period. Chilling requirement is 100 -200 hr below 7°C.
3. Non-astringent cultivars require warm conditions for fruit maturation than the astringent type.

Propagation

Seeds, cuttings, grafting, budding

Rootstocks

D. kaki, D. lotus and *D. virginiana*

Planting and Spacing

Best planting season is July or early August.

5.0 × 2.5 m and 5.0 × 3.0 m

Training and Pruning

Modified central leader system

Flowering, Pollination and Fruit Set

Flowers are produced on new growth in spring. Three types of flowers are produced male, female, hermaphrodite in addition the plant be monoecious or dioecious.

Pollination

Few cultivars which bear pistillate and staminate flowers are used as pollinizers.

For example, Garley, Akagak. One pollinizer tree is interplanted with every 8-10 trees of commercial cultivar.

Persimmon exhibits a double sigmoid growth pattern.

Harvesting

Persimmons mature between early autumn winter depending on the cultivar. Fruits are harvested when fully covered and sugar levels have exceeded 14° Brix.

Yield

40kg fruits/tree (30t/ha)

Storage

At 0°C fruits can be stored upto 2 months but under CA stored upto 5-6 months.

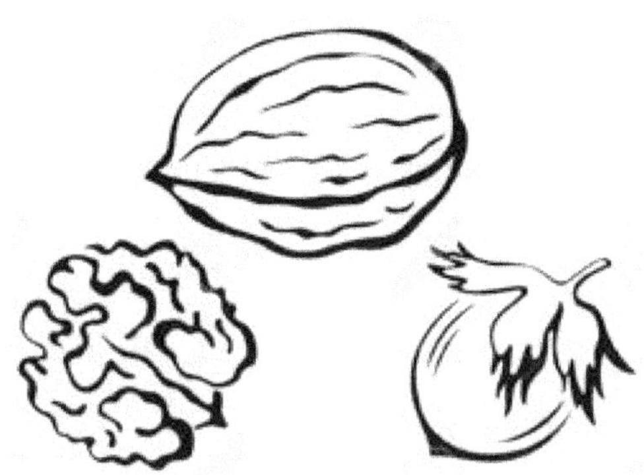

Chapter 13

Walnut

WALNUT (*Juglans sp.*) is the most important temperate nut fruit of the country. It is grown in Jammu & Kashmir, Uttar Pradesh and Himachal Pradesh.

Walnut belongs to family Juglandaceae and genus Juglans. The genus Juglans includes about twenty one species among which *Juglans regia* is the most important, being commercially cultivated in many countries. The somatic chromosome number (2n) of *Juglans* species is 32. The origin of the word 'walnut' has been traced to 'Gaulnut'. The nut is also called as 'Persian walnut' 'English walnut' or 'Carpatian walnut', but the use of term 'Persian walnut' has been considered more appropriate for the nut, because it indicates the place of its origin.

Uses

1. Walnut is the richest source of Vitamin – B_6.
2. It contains a good amount of various vitamins of 'B' group.
3. Walnut is the richest source of fat (64.0 g).
4. The nuts are processed by blanching, slicing, chopping, grinding into paste and butter and combining with other foods.

Climate

The plant can tolerate as low as -11°C during deep dormancy without serious damage but as soon as growth commences after dormancy, the temperature even two or three degrees below freezing point (0°C) kills leaves, shoots and flowers. The chilling hour requirement varies with the cultivars as 700 and 1500 hours are needed for Payne and Franquette cultivars, respectively, at or below 7°C. High temperature greater than 38° C results in blank nuts.

Rainfall

An annual rainfall of about 80 cm is considered sufficient for the cultivation of walnut.

Soil

The top soil should be fertile and well drained.

Rootstocks

Tolerant to Chlorosis

Juglans microcarpa

Adaptability to Severe Cold

Juglans mandschurica, J. cinerea, and *J. nigra.*

J. regia is tolerant to black line disease and *J. hindsii* is resistant to root knot nematode.

Paradox: *Juglans hindsii* × *Juglans regia.* It is tolerant to Apoplexy.

Royal: *Juglans hindsii* × *Juglans nigra*

Vilmorin: *Juglans regia* × *Juglans nigra*

Propagation

Walnut can be propagated either by seed or by vegetative methods.

Vegetative Propagation

Grafting, budding, stooling and micropropagation.

Planting

The seedlings should be planted 10 m apart, while for budded or grafted plants it should be 7 × 7 or 8 × 8 m apart. The ideal time for planting is December–March.

Training and Pruning

Modified central leader system is most ideal for training since it provides very good strength to its framework.

Flowering

Walnut is a monoecious plant. Staminate flowers emerge laterally on the previous seasons growth on catkins. Dichogamy is a rule in walnut.

Cultivars

1. Hartley
2. **Payne:** Dichogamy is absent. It can be grown in solid blocks because pollen and stigma mature simultaneously.

3. Franquette
4. Serr
5. Ashley
6. Sunland
7. Howard
8. Sejnova

Harvesting and Postharvest management

1. Walnuts are usually harvested when hull colour changes from green to yellowish with cracks or when splitting starts at suture from pedicel end.
2. Nuts should be harvested at PTB stage (when packing tissue turns brown).
3. After harvesting, nuts should be properly dehulled, washed and dried.

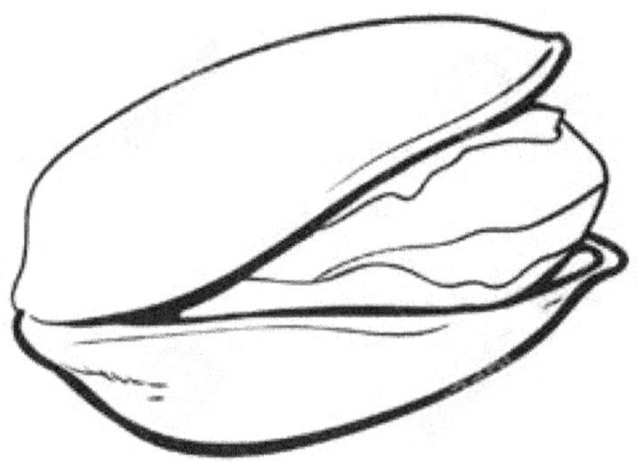

Chapter 14

Pistachio Nut

Botanical Name*:* *Pistachia vera*

FAMILY: Anacaradiacea

ORIGIN: Central Asia

Uses

1. Pistachia can also be used to prepare marmalade (marmalade–dried product produced from pulp of citrus)
2. Sucrose is the main sugar present.
3. Anthocyanin pigment-cyanidine-3-β-galactoside.

Propagation

Seed, T-budding, Shield budding, Veneer grafting and Micro grafting.

Training System

Modified central leader system.

Varieties

Kerman, Joley, Fax, Lassen.

Pollinizers

1. **Peters:** Universal pollinizer produces abundant pollen.
2. Chiko

Climate

1. 700 hrs chilling below 7° C is required.
2. Can with stand -30° C during winter and 42.2° C during summer.

Rootstocks

P. altantica

P. integrima

P. terchinthus

These rootstocks are resistant to root knot nematode and verticillium wilt.

Xenia and metaxenia are seen in pistachio nut

1. Blank seedless fruit production is widely prevalent in all species of pistachio nut.
2. Endocarp dehiscence – When the nut matures the shell covering endocarp separates easily from shell which partially exposes the kernel at apical end.

Chapter 15

Pecan Nut

Pecan nut is a valuable **Horticulture gift of North America** to the world occupying 5th rank among leading tree nuts because of its excellent nutty flavour and superb tree characteristics. In USA, Pecan is considered as **Queen of nuts** because of its value both as a wild and as a cultivated nut.

Botanical Name: *Carya illinoensis*

Family: Juglandaceae

2n=32 **origin:** North America

Besides USA, the worldwide distribution of Pecan is confined to Australia, Canada, Egypt, India, Israel, Mexico, Peru, Turkey and South Africa. All the species of the genus Carya are together know as Hickories and the natural hybrid between Pecan and other are known as HicansPecan nut is a rich source of Fat, protein, carbohydrates and minerals.

Uses

Traditionally pecan is considered as highly nutritive and energy giving nut fruit which is usually taken roasted or salted to supplement normal diet. Over 90 percent of are sold shelled and rest in shell. Nuts are usually used to add aroma flavour crispness, mealiness, tenderness, a rich colour, or to garnish a large number of dishes. The greatest use of Pecan is in baking products such as fruit cakes, custard pies, cookies, nut bread a decision cake fillings and ice cream. Pecan tree is an exceedingly beautiful plant which can be used as shade and ornamental plant.

Cultivars

Stuart, Western, Desirable, Wichita, Cheyenne, Mohawk, Mahan, Nellice, Burkett, Pawnee, Moreland, Ukulinga, Chickasaw.

Climate

Pecan grows in warm temperature climate. It requires 240-280 days growing season with a mean temperature at about 26.7°C. Mean temperature of 3 coldest months should be between 7.2°C and 12.8°C and with at least 400 hrs chilling temperature at or below 7.2°C.

Soil

Sandy loam well fertile soil, rich in organic matter deep, well aerated, well drained soil with pH of 6.4 is ideal.

Propagation

Seed and vegetative propagation is done by cutting, stooling, budding, grafting, *etc.*

Planting

10m × 10m

Training and Pruning

Pecans are generally trained in central leader system.

Manuring and Fertilizers

It has been recommended that pecan require 100 kg of Farmyard manure every year in December in addition apply 500 g NPK mixture per gear increasing with the age of tree and 8kg of mixture to the fully bearing trees of 16 years and above.

Flowering

Pecan starts fruit production in 2 to 6 years depending on cultivar and vegetative growth. The staminate and pistillate flowers are organized on same tree and clustered to catkin and spike respectively. Pecan is mostly cross-pollinated and self-pollination is very rare. Wind is a pollinating agent and carries pollen to about 900m. The planting of 5 percent pollinizers has been recommended to cause adequate pollination.

Fruit Growth and Development

After pollination fruit does not grow for about 45 days and slowly for the next 15 days and thereafter enlarges rapidly for about 30 days reaching almost its final volume. The development of colour marking as dots and streaks of surface of shell is an indication of nut maturation.

Harvesting

Generally nuts are thrashed down with bamboo poles or allowed to shed naturally on ground where these are gathered manually.

Yield

Yield ranging from 200-400 kg per tree is expected in fully grown trees.

Drying and Storage

Pecans are dried soon after harvest to bring the moisture level of kernels to around 4.5 per cent.

Chapter 16

Hazel Nut

SCIENTIFIC NAME: *Corylus avellana*

FAMILY: Corylaceae/Betulaceae

Chromosome No. X=11

Origin

Europe and Asia Minor

Kernal

Oil content ranges from 60.58 per cent to 64.62 per cent.

Turkey is the world's leading producer of filberts. The other major producing countries are **Italy**, **Spain**, **and USA**. Another species *Corylus maxima* may have originated from the South-eastern Europe-western part of Asia. The cobnuts might be the result of natural selection and hybridization of these two species.

Cobnuts = *Corylus avellana* x *Corylus maxima*

Climate and Soil

Long cool summer is favourable for the development of the nuts than long warm summer and a rather long exposure to moderate cold is necessary to break the rest period. Neutral to slightly acidic soils having about pH 6 are favourable for filbert culture.

Varieties

Barcelona, Butler, Northampton, Ana, Self Husker, Skinner, Tondagentile, Tomanu.

Rootstocks

C. colurna is recommended because of its vigour, extensive root system and non-suckering habit.

Propagation

Cuttings, layering, grafting and budding.

Planting

Trees should be spaced 5-7m apart.

Open centre system of training.

Flowering

Flowering pattern is monoecious. The pistillate flowers (female) which develop into the nuts are usually borne in lateral and terminal buds. In the previous season's growth, the staminate flowers (male) which provide pollen are borne in naked catkins in short stalks either laterally in the previous season growth or terminally on shoot spurs which may also bears leaf or fruit bud.

Pollination

Self-incompatibility in filbert cultivars is common which makes interplanting with pollinizer essential in order to provide cross pollination. The blooming season under normal condition extends over a period of three months.

Fruit Growth

The shell is formed from the ovary, *i.e.,* the ovary wall where as the kernel is formed from the embryo.

Harvesting

Fully mature fruits of filberts drop on the ground. Nuts are gathered 2-3 times during the season.

Yield

The filberts vary in yields according to the age and vigour of trees. The yield may be up to 910 kg/acre.

Storage

After the filberts are dried to an in-shell moisture content of 7.8 per cent, the nut should be graded, fumigated and sealed in plastic lined boxes.

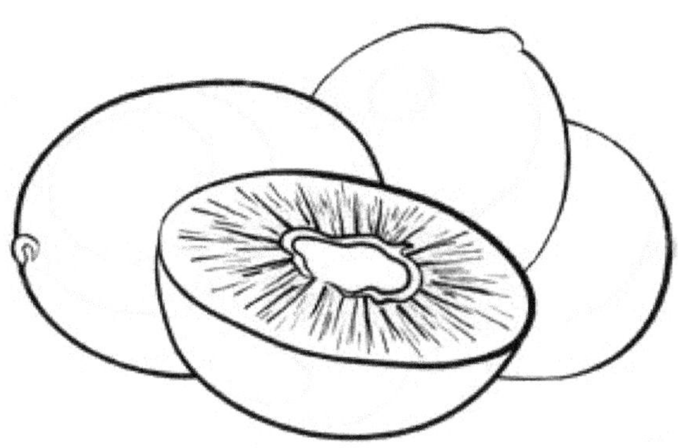

Chapter 17

Kiwi

Kiwi is known as 'China's miracle fruit' and the horticultural wonder of New Zealand. The kiwi fruit or Chinese goose berry (*Actinidia deliciosa*) has gained enormous popularity in the past two decades in many countries of the world. It is placed in the family Actinidiaceae (Dilleniaceae). Its chromosome number is 2n = 58. The kiwi fruit is native to China. The kiwi is a Auto hexaploid.

Soft haired kiwi fruit: *Actinidia chinensis* var. *chinensis*

Stiff haired kiwi fruit: *Actinidia chinensis* var. *hispida*

Spiny kiwi fruit: *Actinidia chinensis* var. *setosa*

Acidity in kiwi fruit is attributable to the presence of malic acid, citric acid and quinic acid.

Pleasant aroma in kiwi fruit is due to Ethyl butanoate, Hexanol and Trans-hex-2-enal.

Kiwi is functionally dioecious and a deciduous fruiting vine.

Actinidin (Cysteine protease) is a thiol protease enzyme present in kiwi fruit.

Soil and Climate

Deep well drained, sandy-loam soil with good amount of organic matter is ideal for its cultivation. A soil pH 5.5-6.5 is ideal for vine growth and fruit production. It requires 600-800 chilling hours to break dormancy.

Planting

Planting is done during December to January.

Important Pistillate Cultivars

1. Abbott
2. Allison
3. Bruno
4. Hayward
5. Monty
6. Greensil
7. Top Star: It is a bud mutant of Hayward.
8. Green Light: It is a bud mutant of Hayward.

Important Staminate Cultivars

1. Tomuri and Matua. Tomuri is a good pollinizer for Hayward.

Propagation

Seed, Soft wood cuttings, grafting, budding, hard wood cuttings and stem cuttings.

For pollination, 1 staminate plant is provided with 6 pistillate plants.

Rootstocks

1. *Actinia arguta*
2. *Actinidia kolomikta*

Training

T-bars

In kiwi, the fruits are borne on the current season's growth arising from the bud developed in the previous year. Kiwi fruit requires high amount of chlorine because chlorine deficiency adversely affects the growth of shoots and roots. Kiwi fruit does not have a long juvenile period, the flowering begins after 2 or 3 years after planting. Kiwi fruit does not follow typical respiration pattern of either climacteric or non – climacteric pattern. The kiwi fruit exhibits triple sigmoid growth curve. In controlled atmosphere conditions, kiwi fruit can be stored for more than six months.

Maturity Index

TSS – 6.2 per cent

Storage Disorders

1. White core
2. Translucency
3. Graininess
4. **Internal breakdown:** Ripening temperature more than 20°C causes internal breakdown.

Index

Y

www.ingramcontent.com/pod-product-compliance
Ingram Content Group UK Ltd.
Pitfield, Milton Keynes, MK11 3LW, UK
UKHW021952270726
14060UKWH00002B/479

9 789390 371716